3 Crystals

Growth, Properties, and Applications

III-V Semiconductors

Springer-Verlag
Berlin Heidelberg New York 1980

Managing Editor

Priv.-Doz. Dr. H. C. Freyhardt
Kristall-Labor der Physikalischen Institute
Lotzestr.16–18
D-3400 Göttingen

and

Institut für Metallphysik der Universität Göttingen
Hospitalstr. 12
D-3400 Göttingen

ISBN-13: 978-3-642-67613-0 e-ISBN-13: 978-3-642-67611-6
DOI: 10.1007/978-3-642-67611-6

Softcover reprint of the hardcover 1st edition 1980

2152/3140-543210

Foreword

Springer-Verlag, Berlin Heidelberg, in conjunction with Springer-Verlag New York, is pleased to announce a new series:

CRYSTALS

Growth, Properties, and Applications

The series presents critical reviews of recent developments in the field of crystal growth, properties, and applications.

A substantial portion of the new series will be devoted to the theory, mechanisms, and techniques of crystal growth. Occasionally, clear, concise, complete, and tested instructions for growing crystals will be published, particularly in the case of methods and procedures that promise to have general applicability.

Responding to the ever-increasing need for crystal substances in research and industry, appropriate space will be devoted to methods of crystal characterization and analysis in the broadest sense, even though reproducible results may be expected only when structures, microstructures, and composition are really known.

Relations among procedures, properties, and the morphology of crystals will also be treated with reference to specific aspects of their practical application. In this way the series will bridge the gaps between the needs of research and industry, the possibilities and limitations of crystal growth, and the properties of crystals.

Reports on the broad spectrum of new applications – in electronics, laser technology, and nonlinear optics, to name only a few – will be of interest not only to industry and technology, but to wider areas of applied physics as well and to solid state physics in particular.

In response to the growing interest in and importance of organic crystals and polymers, they will also be treated.

Scientists who have agreed to participate in the new series are:
Prof. Arizumi, Nagoya, Dr. Bardsley, Great Malvern, Prof. Bethge, Halle, Prof. Chernov, Moscow, Dr. Freyhardt, Göttingen, Dr. J. Grabmaier, München, Prof. Haussühl, Köln, Prof. Hoppe, Gießen, Prof. Kern, Marseille, Dr. Laudise, Murray Hill, Prof. Nitsche, Freiburg, Prof. Rabenau, Stuttgart, Dr. Rooijmans, Eindhoven, Prof. White, University Park, Prof. Witt, Cambridge, and Dr. Young, Jr., Oak Ridge.

Managing editor for Volume 1 was Dr. C. J. M. Rooijmans, Philips' Gloeilampenfabrieken Eindhoven in The Netherlands. Managing editor for Volume 2 and all subsequent volumes is Dr. Freyhardt, Göttingen.

Two 200 to 250 page volumes are planned annually.

The Publisher | The Editors

Units

1 N *(Newton)* = 1 kg m/s^2 = 10^5 dyn
1 kp = 9,81 N ≈ 1 da N
1 Pa *(Pascal)* = 1 N/m^2 = 10^{-5} bar
1 at = 1 kp/cm^2 = 0,981 · 10^5 Pa =
= 0,981 da N/cm^2
1 atm = 760 Torr = 1,013 · 10^5 Pa
1 Torr = 133,323 Pa

1 J *(Joule)* = 1 Nm = 1 Ws = 10^7 erg
1 kWh = 3,6 MJ = 559,8 kcal
1 eV = 1,6021 · 10^{-19} J
1 PSh = 2,648 · 10^6 J = 0,7355 kWh
1 kpm = 9,81 J = 2,724 · 10^{-6} kWh
1 kcal = 4,1868 · 10^3 J = 1,163 · 10^{-3} kWh

1 kcal/kg K = 4,187 · 10^3 J/kg K
1 kWh/kg K = 3,6 · 10^6 J/kg K = 3,6 MJ/kg K

1 kcal/m^2 h = 1,163 W/m^2
= 11,63 · 10^{-8} kW/cm^2
1 cal/cm^2 s = 4,187 W/cm^2

1 kcal/m h K = 0,01163 W/cm K
1 cal/cm s K = 4,187 W/cm K

1 kcal/m^2 h K = 1,163 W/m^2 K
= 11,63 · 10^{-8} kW/cm^2 K
1 cal/cm^2 s K = 4,187 W/cm^2 K

1 J/s = 1 Nm/s = 1 W = 1 VA = 10^7 erg/s
1 PS = 0,735 kW = 632,5 kcal/h
1 kpm/s = 9,81 Nm/s = 9,81 · 10^{-3} kW
1 kcal/s = 4,187 kW = 5,692 PS
1 kcal/h = 1,163 J/s = 1,163 · 10^{-3} kW

1 V *(Volt)* = 1 W/A	1 F *(Farad)* = 1 C/V
1 Ω *(Ohm)* = 1 V/A	1 Wb *(Weber)* = 1 Vs
1 S *(Siemens)* = 1/Ω	1 T *(Tesla)* = 1 Wb/m^2
1 C *(Coulomb)* = 1 As	1 H *(Henry)* = 1 Wb/A

1 Pa s = 1 Ns/m^2 = 1 kg/ms
1 P *(Poise)* = 10^{-1} Pa s, 1 da P = 1 Pa s
1 kp s/m^2 = 9,81 Pa s
1 kp h/m^2 = 0,0353 · 10^6 Pa s = 0,0353 MPa s
1 St *(Stokes)* = 10^{-4} m^2/s
1 m^2/h = 2,778 · 10^{-4} m^2/s

1 lm *(Lumen)* = 1 cd sr
1 lx *(Lux)* = 1 lm/m^2
1 sb *(Stilb)* = 10^4 cd/m^2

1 Ci *(Curie)* = 3,7 · 10^{10}/s
1 rd *(rad)* = 10^{-2} J/kg
1 rem = 10^{-2} J/kg
1 R *(Röntgen)* = 258 · 10^{-6} C/kg

Contents

Editorial Board

Growth of Binary III-V Semiconductors from Metallic Solutions

Klaus-Werner Benz[1] and Elisabeth Bauser[2]

[1] Physikalisches Institut, Kristallabor, Universität Stuttgart, D-7000 Stuttgart 80, Federal Republic of Germany

[2] Max-Planck-Institut für Festkörperforschung, D-7000 Stuttgart 80, Federal Republic of Germany

Solution growth of binary III-V-semiconductors has proved to be an important crystal growth technique for the preparation of highly perfect and well-defined crystals. The aim of this paper is to review solution growth methods of bulk single crystals, combined with a review of selected aspects of thin-layer single-crystal growth, which are regarded to be equally relevant for bulk solution growth: Growth mechanisms in correlation with growth conditions in liquid phase epitaxy are discussed. The resulting properties of the thin layers, such as surface morphology and dopant homogeneity, are described.

Table of Content

1 Introduction

Since the discovery and development of the semiconducting properties of III–V compounds by Welker in 1952[1)], these materials have gained importance for the production of electronic devices, e. g., light-emitting diodes (LED's), lasers, microwave devices, galvanomagnetic devices, etc. Single crystals with a high degree of perfection and well-defined behavior (e. g., high purity, controlled doping, etc.) are needed as basic materials for such applications.

Table 1. Binary III-V-compounds with some selected properties

above the compound	:	bandgap (300K) [eV]
below the compound	:	lattice constant [Å]
below the element	:	covalent tetrahedral radii [Å]
hatched area	:	important binary compounds as basic materials for device applications

	N	P r = 1,127	As 1,225	Sb 1,405
B	BN (2,51)	BP (4,538)	BAs (4,777)	BSb
AL 1,229	ALN (3,111)	(2,45) ALP (5,4625)	(2,14) AlAs (5,6611)	(1,62) AlSb (6,1355)
Ga 1,225	(3,39) GaN (3,180)	(2,26) GaP (5,4495)	(1,43) GaAs (5,6419)	(0,70) GaSb (6,094)
In 1,405	InN (3,533)	(1,35) InP (5,868)	(0,356) InAs (6,058)	(0,180) InSb (6,478)

Table 1 depicts 16 III–V compounds with semiconducting properties. The most important binary III–V semiconductors are GaAs, AlAs, GaP, InP with a bandgap exceeding 1 eV, and InAs and InSb with a bandgap below 1 eV.

Most of the III–V compounds have a Zinc blende structure. Exceptions are the nitrides (wurzite structure) and the cubic modification of BN. The Zinc blende structure consists of two interpenetrating f. c. c. sub-lattices, one describing the positions of the atoms of the III-component (O,O,O,f.c.), the other, atoms of the V-component (1/4, 1/4, 1/4, f.c.). Each atom is located at the center of a regular tetrahedron of unlike atoms (4 atoms per unit cell). This tetrahedral structure implies that the lattice does not reveal the center of an inversion symmetry, and these crystals therefore exhibit a polar behavior in the <111> direction of the III component alternating with atoms of the V component, but with unequal spacing. Thus, two distinct (111) surfaces, a (111) A and a (111) B, can occur.

Solid-liquid phase diagrams must be considered in order to understand the growth process from metallic solutions, and to be able to obtain reproducible growth results. For example, Fig. 1 shows the phase diagrams of GaSb and InSb, which are representative of the other III–V compounds.

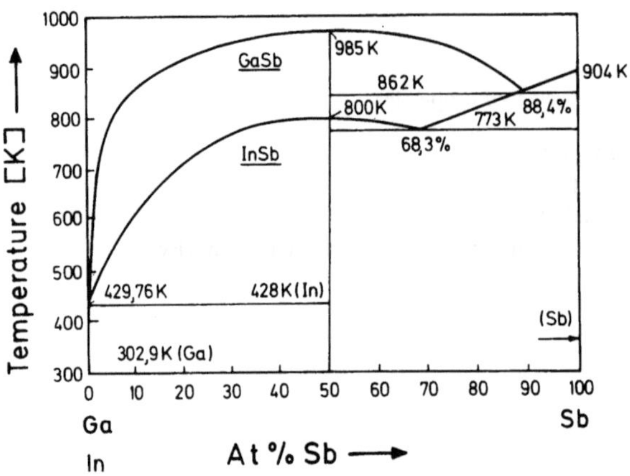

Fig. 1. Typical phase diagrams of III-V semiconductors: GaSb and InSb

From a technological point of view, all III–V semiconductors can be divided into three main groups with respect to the vapor pressure of the V component:

- low vapor pressure at the melting point (10^{-4}–10^{-1} bar), GaSb, InSb
- medium vapor pressure at the melting point (~ 1 bar) GaAs, InAs
- high vapor pressure at the melting point (> 10 bar) InP, GaP.

Problems in growth technology arise in melt growth methods, as well as in solution growth methods with increasing vapor pressure of the compounds. The following crystallization methods are applied in typical growth technologies of III–V semiconductors:

- melt growth: Czochralski, Bridgman, gradient freeze (with liquid incapsulation variants for compounds with high vapor pressure), zone melting (mainly for raw material production), floating zone method (only for diameters up to 8–10 mm)
- solution growth:
  thin films: Liquid Phase Epitaxy,
  bulk crystals: to a small extent: Czochralski growth, Bridgman method
  others: Traveling Solvent Techniques; Solute Synthesis Diffusion (Temperature Gradient Growth)
- vapor growth: (thin films)
  Vapor Phase Epitaxy
  Molecular Beam Epitaxy

Solution growth of III–V semiconductors has some major advantages over melt growth methods:

a) no or reduced vapor pressure problems due to the lower growth temperature,
b) reduced attack of construction materials (crucibles and reactor materials),
c) the lower growth temperature reduces the deviation from stoichiometry by purely thermodynamic reasons, which results in more perfect crystals.

Figure 2 shows the region of existence of solid GaAs as a function of the temperature calculated by Logan and Hurle[2]. In another example, GaSb grown from the melt always exhibits a p-type behavior ($p \sim 2 \cdot 10^{17}$ cm^{-3}) due to the native defect $[V_{6a}]$ $[Ga_{Sb}]$[3].

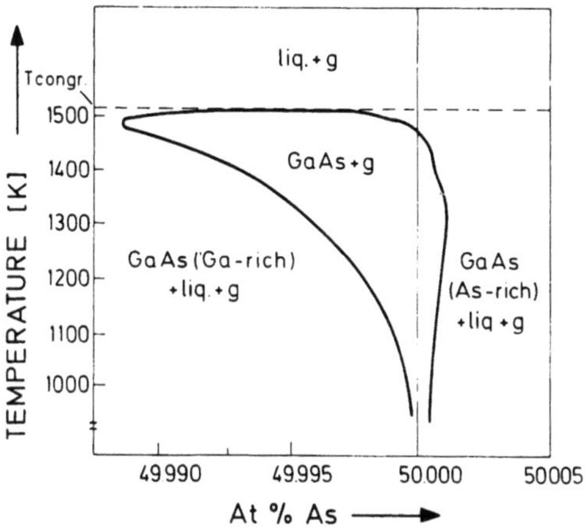

Fig. 2. Existence region for GaAs, calculated by Logan and Hurle[2]

Growth from dilute solutions is therefore extensively used to facilitate the preparation of the III–V compounds at temperatures for below their nominal melting points.

The following article describes various solution growth methods for III–V semiconductors. Bulk methods can be regarded as an effort to produce high-quality substrate materials for device application. Compared with Liquid Phase Epitaxy, only a few articles are available on bulk solution growth. In chapter 2 an overview on experimental arrangements in connection with various growth results will be presented.

Since excellent review articles (references, see Chapt. 3.1) are available for Liquid Phase Epitaxy, a few selected subjects, which have not yet been covered, will be discussed in this paper. In order to get a better understanding of III–V semiconductor solution growth, chapter 3 will deal with solution growth mechanisms which have been extensively studied for Liquid Phase Epitaxy.

2 Bulk Crystal Growth Methods

2.1 Preliminary Remarks

Methods for the growth of bulk crystals from solution have not often been used in the past. The reason is the low pulling rate, which must be applied when leaving the stoichiometric composition of the melt. In the solution of a metallic group-III solvent, the transport of the group-V solute is mainly governed by diffusion processes with low values of diffusion coefficients. The growth rate decreases drastically from several mm/h to several mm/day, at which time the stoichiometric melt is replaced by a (nonstoichiometric) solution. A further difficulty is found in the increasing tendency of the growth interface to become unstable[4].

In this chapter we will give a survey on various growth methods which enable the reproduction of crystals of welldefined shape and high quality.

2.2 Application of Czochralski and Bridgman Techniques in Solution Growth

Only a few studies have been carried out by Czochralski-, Bridgman- or gradient freeze techniques, as compared with the numerous melt and LPE growth experiments. It has generally been difficult to control and program temperatures and mechanical movement in such a way as to avoid solvent inclusions. As a result, the crystallinity of such macro-crystals is only poor.

Various attempts have been undertaken to grow GaSb (by Czochralski) and modified Bridgman techniques from non-stoichiometric melts (Ga-rich and Sb-rich solutions)[3, 5–7] in order to reduce the native defect concentration. The size of the GaSb single-crystalline pieces ranged from several mm^3 to about 1 cm^3. The residual acceptor density in GaSb could be reduced down to $p \cong 3 \cdot 10^{16}$ cm^{-3}.

Recently, InP crystals have been grown by a gradient freeze technique from In-rich solutions[8]. A horizontal Bridgman system consisted of a high-temperature furnace controlling the In temperature at 940 °C and a low-temperature furnace controlling the partial pressure of P at 420 °C. The system was capable of producing a 500 g ingot per run; the applied growth rate was about 10 mm/day. Hall measurements in single crystal grains showed that the first-to-freeze portion of the crystal had a residual carrier concentration of $6{,}8 \cdot 10^{14}$ cm^{-3} and a liquid nitrogen mobility of 91 000 $cm^2 V^{-1} s^{-1}$. In addition to these measurements, low temperature photoluminescence studies indicated the purity and crystal perfection of these grains as equivalent to that of LPE and VPE layers[9].

In summary, we conclude that Czochralski and Bridgman techniques based on the use of metallic solutions can be employed III-V compounds; however, many growth attempts are necessary in order to increase the size of such crystals.

2.3 Travelling Solvent Techniques

The Travelling Solvent Techniques can be devided into two main parts, the Travelling Solvent Method (TSM) and the Travelling Heater Method (THM). The first one was introduced by Pfann[10], who called it Temperature Gradient Zone Melting (TGZM). However, in Pfann's original meaning a melt, and not a solution, was used for TGZM. THM for III–V compounds was first introduced by Broder and Wolff.[11] They grew GaP from a Ga solution at a growth temperature of about 1000 °C. In TSM no moving parts are necessary, whereas in THM either a moving ampoule or a travelling heater can be used. A review of travelling solvent techniques up to 1973 was given by Wolff and Mlavsky[12].

In III–V semiconductor growth and technology, TSM has sometimes been used in the preparation of small-area well-defined crystals and large-area abrupt-rectifying p-n junctions. On the other hand, THM is better suited to the growth of large single crystals of well-defined form. The next two chapters deal with some of the TSM and THM growth results.

2.3.1 The Travelling Solvent Method (TSM)

The principle of TSM is shown schematically in Fig. 3. A thin solution zone passes through a crystal when a strong temperature gradient[13] is applied. The lower crystal in Fig. 3 acts as a single crystalline seed; the upper feed material can be polycrystal-

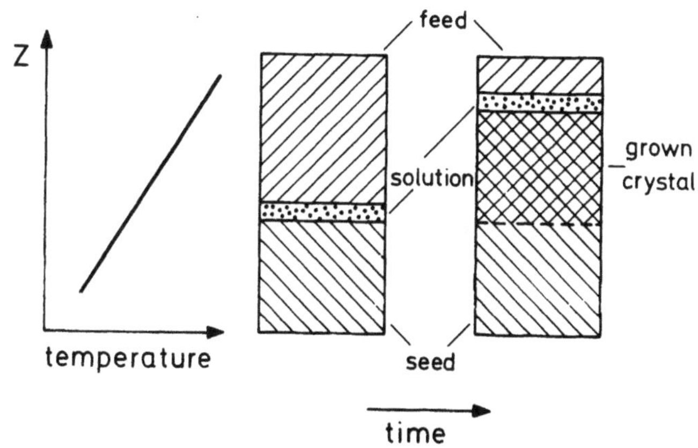

Fig. 3. Travelling Solvent Method (TSM) – schematically

line. The seed/solution boundary advances through deposition of material at the cooler interface. This material is dissolved from the receding feed at the higher temperature interface with the liquid zone.

The migration of a solution zone under the influence of a temperature gradient has been analyzed theoretically by Tiller[14]. The travelling rate of the solution zone is proportional to the temperature gradient and depends on thermal conductivity of the solid material and the liquid, as well as upon the atomic kinetics of the freezing and melting process. If $v.L./D \ll 1$ (v = travelling rate of the liquid zone, L = zone length, D = diffusion coefficient), the influence of diffusion in the zone can be disregarded.

A series of III–V compounds, e. g., GaAs[13], GaP[15], $GaAs_yP_{1-x}$[16], GaSb[17], $Ga_xIn_{1-x}P$[18], InAs[12], has been grown by TSM. GaAs growth experiments have been done at about 900 °C with a temperature gradient of 60–80 deg/mm[13]. The zone-moving equipment consisted of a carbon strip heater (compare Fig. 4) housed in

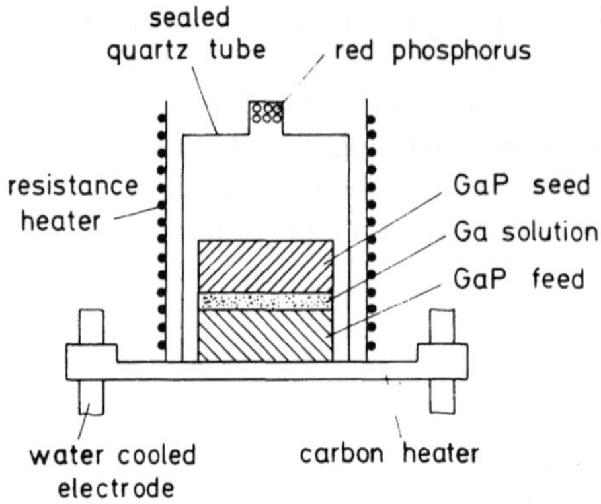

Fig. 4. Schematic drawing of a TSM-growth apparatus for GaP[15)]

a bell jar which could be evacuated and back-filled with inert gas. A sandwich structure consisting of GaAs seed and feed material with a thin layer of Ga in between (25 μm thick) was placed beneath the heater (above the heater in the case of GaP[15)], see Fig. 4). The resultant growth rates were about 0,5 mm/h (12 mm/day); the crystal had a diameter of 6 mm with a grown length up to 6 mm. In the same way GaP crystals have also been grown by TSM (Fig. 4) at average temperatures as low as 850 °C[15)]. In a 2,5 mm sandwich (1,25 mm thick slices of GaP and a Ga solution zone of 25 μm), a temperature gradient of about 200 °C was necessary before significant zone travelling occurred.

An interesting result of TSM growth is the fact that the dislocation density in the grown crystal can be considerably lower than in the seed crystal. The reduction of dislocation density has been demonstrated in GaAs[13, 19)] and GaSb[17)] by using etching techniques as well as x-ray characterization methods. In GaAs TSM-crystals, Weinstein et al.[19)] found a decrease in dislocation density from $8 \cdot 10^4$ cm^{-2} (seed) to $1 \cdot 10^4$cm^{-2} (grown crystal). The dislocation elimination appeared to occur within the first 150–200 μm of growth. In the case of a planar growth interface, the mechanism of dislocation removal after the beginning of TSM growth can be explained by either a closed loop formation, or by decrease through a transformation to other types of dislocation which then propagate alongside to the crystal surface[17)].

2.3.2 The Travelling Heater Method (THM)

The growth of large crystals is formed by the Travelling Heater Method (THM). Figure 5 shows the principle THM arrangement. The solvent zone, which is placed between a solid seed and the feed material, is heated by a special heater (electrical resistance heater[20)], direct RF[21)] coupling, laser heating or mirror furnace facility[22)]). By moving the charge with respect to the fixed heater (Fig. 5), crystal-

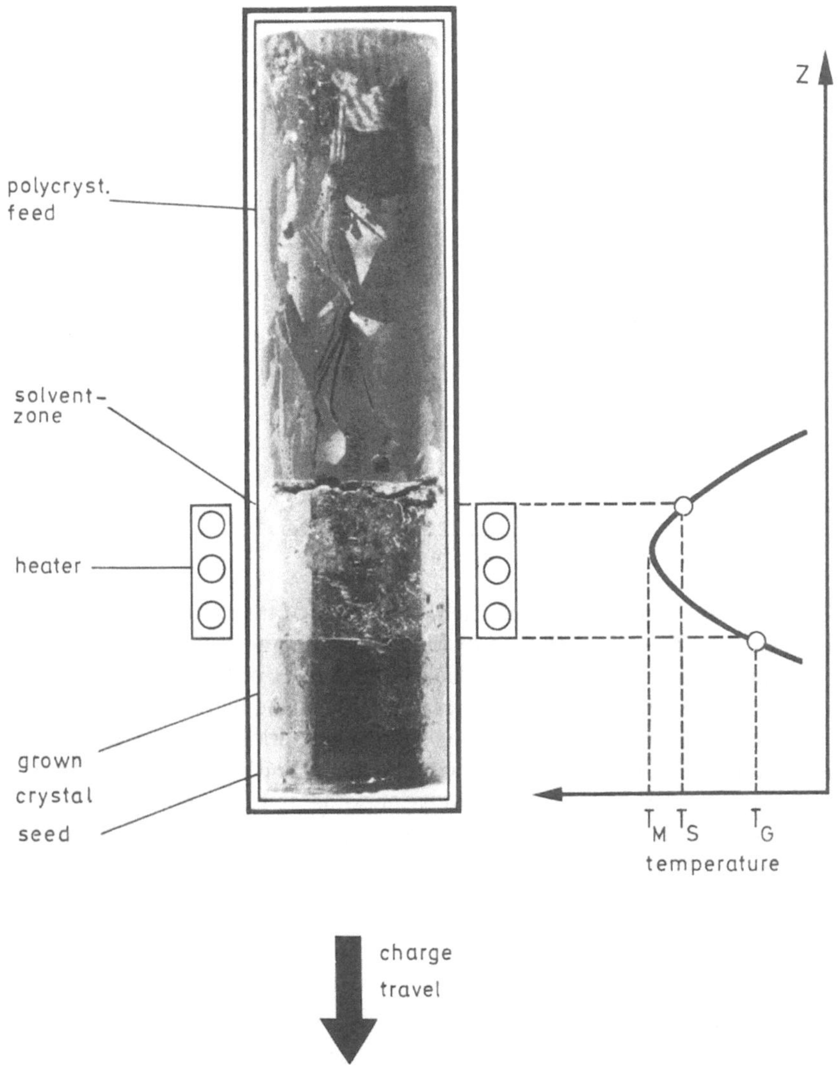

Fig. 5. Principle of the Travelling Heater Method (THM), vertical arrangement. Photograph of an "as grown" GaSb sample. Solution temperature T_M = 560 °C, growth rate v = 2,2 mm/day, crystal diameter: 10 mm

lization at the advancing seed/solvent interface can be achieved. Simultaneously, feed material is dissolved at the feed/solvent phase boundary. The solute transport in the liquid zone may be established by diffusion and/or convection.

THM growth has been applied to various III–V compounds, e. g., GaAs[23, 24], GaP[11, 21] GaSb[25, 26] InSb[25, 26] and InP[26], as well as ternary compounds[23].

Typical growth data for these materials are summarized in Table 2.

Table 2. THM-Growth parameters of various III-V compounds

III-V Semiconductor	Solvent	Growth temperature °C	Growth rate mm/day	References
GaAs	Ga	1000	⩽ 5	Hemmat et al.[23]
	Ga	900	2	Wilcox[24]
GaAsP	Ga	1000	⩽ 5 to 6	Broder and Wolff[12]
GaP	Ga	1000–1100	⩽ 5 to 6	Broder and Wolff[12]
	Ga	1160	4	Plaskett et al.[21]
GaAlAs	Al-Ga	1000	⩽ 3 to 5	Hemmat et al.[23]
GaInP	In-Ga		⩽ 3 to 5	Hemmat et al.[23]
GaSb	Ga	500	2,5 ± 0,5	Benz and Müller[25]
InSb	In	400	2,5 ± 0,5	Benz and Müller[25]
InP	In	800	< 2,0	Benz and Müller[26]

Before a growth experiment is conducted, knowledge of the axial and radial temperature distribution during crystal growth might be very helpful. This is because the shape of the seed/solution interface influences the quality of the growing crystal. Axial and radial temperature distributions have been calculated by Bell[27] using the numerical integration of the Laplace equation and the cylindrical geometry shown in Fig. 5. In this calculation the boundary between the liquid and the solid, which have different physical properties, was assumed to be an isotherm. Isotherms resulting from this calculation are plotted in Fig. 6. The calculations were done for CdTe as

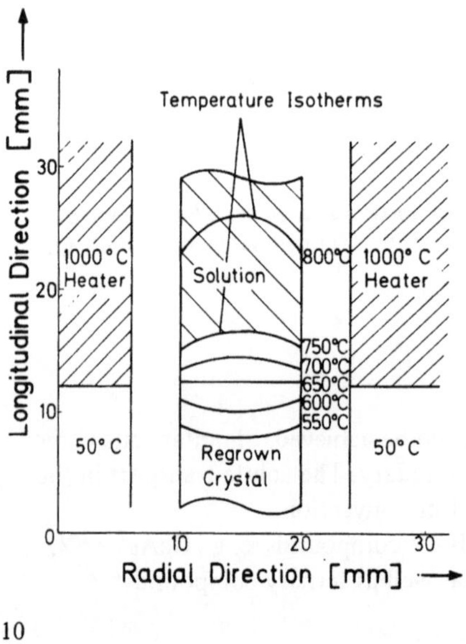

Fig. 6. Travelling Heater Method. Plot of isotherms for a crystal/solution boundary temperature of 750 °C[27]. Crystal: CdTe, solvent Te, heater temperature: 1000 °C

crystal material and a Te solution zone. The ampoule was 6 cm long (diameter 1 cm), symmetrically located within a 2 cm hot zone at 1000 °C. In order to approximate normal conditions, an abrupt transition from the hot (1000 °C) to the cool (50 °C) zone was used. However, in practice there would be a more or less continuous change which would smooth out the transition. At an interface temperature of 750 °C, Fig. 6 shows the convex shape of a crystal/solution boundary. At a 650 °C interface temperature, a concave shape will result.

Another important parameter is the temperature distribution T(Z) along the center of the ampoule. From this curve the temperature gradient at the seed/solution boundary can be determined. Numerical calculations were made by Bell[27], and an analytical expression for T(Z) within the heater region was formulated by Langbein[28]:

$$T(Z) = T_M - \frac{T_H4 - T_M4}{2\,T_M3}\,[\sinh(\gamma^{1/2} T_M{}^{3/2}(Z + \tfrac{1}{6}\beta Z^2))] \qquad (1)$$

with

$\beta = v \cdot \kappa \cdot \lambda$ and $\gamma = 2\,\sigma\,\epsilon/R \cdot \lambda$
σ = Stephan-Boltzmann-constant.

A temperature profile was calculated for a GaSb-THM experiment using the following data and physical characteristics:

Heater tempe rature	T_H	=	973 K
maximum solution temperature	T_M	=	636 K
growth rate	v	=	$2{,}5 \cdot 10^{-6}$ cms^{-1}
heat capacity of the Ga solution	κ	=	2,3 J K^{-1} cm^{-3} [29]
thermal conductivity of the GA solution	λ	=	$3{,}98 \cdot 10^{-1}$ J s^{-1} $cm^{-1}K^{-1}$ [29]
emissivity	ϵ	=	1
radius of the heater	R	=	0,8 cm

Figure 7 shows the measured temperature distribution T(Z) of two positions in the solution zone, together with the theoretical values which in a practical use fit the experimental data very well[a].

The following practical implications have become apparent from these calculations:

For longer solution zones the seed/solution interface becomes concave, whereas for shorter zones it becomes convex. In order to prevent undesirable seeding during crystal growth, the interface should be convexly curved, especially in the immediate vicinity of the ampoule wall. Therefore, a short solution zone with a long heater is

---

[a] The temperature distribution T(Z) can be measured by means of calibration ampoule[24, 25].

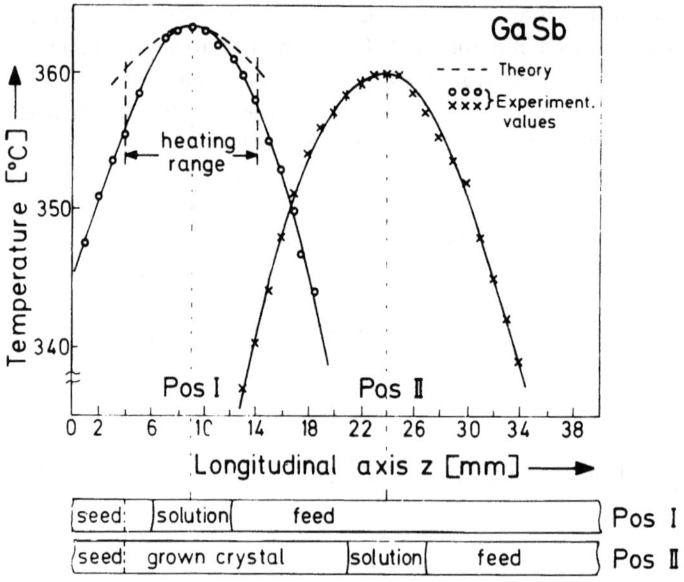

Fig. 7. Temperature distribution T(z) along the longitudinal axis of a THM ampoule (GaSb with Ga solution).
Left diagram: T(z) at the beginning of growth
Right diagram: T(z) after several mm of growth

desirable. A large temperature gradient at the interface, however, implies a short heater (in order to prevent constitutional supercooling and to achieve fast growth rates). Thus, the best configuration must be determined for each different crystal-solution system.

THM Growth Experiments. A series of GaAs crystals have been grown by Wilcox et al.[24, 30] in quartz ampoules using a Ga solution zone and GaAs seed crystals with different orientations. The growth rate was varied between 0,5 mm/day and 20 mm/day, and the temperature of the heater between 850 °C and 1100 °C. The growth of the GaAs crystals was determined by the following parameters: (i) The ratio of the zone length to the length of the heater ℓ/h. The best crystal quality was achieved by ℓ/h greater than 0,5 but smaller than 1. For this study a convex seed/solution interface was observed, which agrees well with the calculations of Bell[27]. (ii) The temperature of the heater. The temperature of the solution zone, which was measured by a special calibration ampoule, was always 60–80 °C lower than the maximum heater temperature. Below the 850 °C heater temperature, not a single crystal could be grown. GaAs crystals with a diameter of 10 mm and several lengths without Ga inclusion have been grown at a heater temperature of 980 °C and a growth rate of about 2 mm/day.
Benz and Müller recently reported[25, 26] on the THM-growth of GaSb and InSb crystals using a vertical system as well as a horizontal one. In the vertical set-up (Fig. 8) a fused quartz ampoule (ID = 10 mm) containing the seed crystal, the solu-

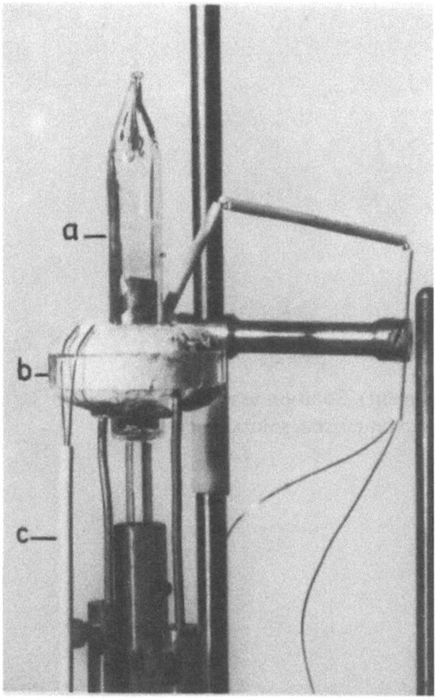

Fig. 8. Vertical THM arrangement with the quartz ampoule and a heating coil *a:* ampoule, *b:* heating coil, *c:* thermocouple

tion, and the polycrystalline feed material was mounted on a travel mechanism with a goniometer, which could be adjusted for an axially symmetric rotation (10 rpm) in the temperature field of the heater. The temperature of the heater was controlled over a period of a week with an accuracy of ± 1 °C.

During preparation of the growth ampoules, seed and feed materials had to be adjusted by grinding and etching the inner diameter of the quartz ampoule (10 mm ± 50 μm) or the quartz bot (horizontal set up) respectively. If the spacing between the seed and quartz ampoule is not small enough, seepage of the liquid solution can occur, often resulting in polycrystalline growth. Prior to growth the cleaned and etched materials have to be put into the growth ampoule by avoiding any contact with outside air[25)].

Various GaSb and InSb crystals were grown in vertical and horizontal growth arrangements. In Fig. 5 and 9, "as grown" samples of GaSb (Fig. 5, vertically grown); diameter: 10 mm, length: up to 30 mm and InSb (Fig. 9, horizontally grown) are shown. The solution temperature T was varied between 410 °C and 550 °C for GaSb. The In-Sb-solution temperature was kept at 400 °C. In order to study the influence of the zone shape on crystal quality, the zone length was varied between 3 and 13 mm in the vertical THM arrangement. The effective heater length was 12 mm. Depending on the zone length, different shapes of the growth interfaces were obtained (Fig. 10). In accordance with the calculations of Bell[27)], the radius r of the interface curvature shrank with decreasing zone length ℓ. For $r < 7$ mm ($\ell < 5$ mm) no single crystal was obtained.

Fig. 9. As grown InSb sample (horizontal THM arrangement). Solution temperature $T_M = 400\,°C$, growth rate v = 4,3 mm/day, from left to right: seed, grown crystal, solution and feed

Fig. 10. Shape of a Ga-Sb-solution zone with an effective zone length of $\ell = 8$ mm

In some cases polycrystalline growth occured (Fig. 11). With zone length $5 < \ell < 8$ mm ($7 < r < 10$ mm), the best growth results (GaSb and InSb) were obtained. In the case of horizontal THM the growth behavior is similar to the results of the vertical THM[25] due to the zone shape. In contrast to the vertical THM, there is no axial symmetry with respect to the growth axis; therefore, the geometry of the zones is more complex.

Fig. 11. Longitudinal section of vertically grown GaSb; seed Te doped. Growth rate v = 2,88 mm/day, polycrystalline growth 6 mm above seed

A similar behavior in the relationship between heater length, zone length, and crystal quality was reported for the THM growth of CdTe[31, 32]. A maximum growth rate is important for GaSb and InSb, as in the case of GaAs – THM growth. The highest growth rate v_{max} which could be achieved in order to get inclusion-free GaSb (T_M = 500 °C) was v_{max} = (2,5 ± 0,5) mm/day. This holds true also for the growth of InSb with T_M = 400 °C. Growth rates of other III–V compounds can be taken from Table 2. As can be seen in Table 2, the growth rates of various III–V compounds are very similar and stand at between 2–5 mm/day. In explaining the value of the maximum growth rate, we can assume a diffusion-controlled solute transport. A relationship between the diffusion coefficient D of the solute in the liquid zone and v_{max} can be calculated by taking the steady-state diffusion equation and the conservation of mass[33]:

$$v_{max} = \left(\frac{D}{\ell}\right) \ln \frac{C_G - C_o}{C_S - C_o} \qquad (2)$$

where ℓ is the zone length and $C_G(C_S)$ the solute concentrations at the growth (dissolving) interface; C_o represents the concentration of V-component in the solid. C_G and C_s are taken from the phase diagram by taking the measured temperatures T_G and T_S from calibration experiments. In the case of vertical THM of GaSb ($T_M \sim 500$ °C), a value of $D \cong 3 \cdot 10^{-4}$ cm^2 s^{-1} can be calculated.

This value is more than one order of magnitude higher than the diffusion coefficients of As and P in liquid Ga at about 500 °C[34] (to date, a D value of Sb in liquid Ga is not available). Therefore, we can assume that in addition to diffusion, the solute transport in the liquid-Ga zone is supported by natural convection. Similar results have been reported for the THM growth of CdTe from a Te solution[31]. In order to avoid contamination and disturbances from the crucible walls, a floating solution zone can be used for the THM-growth of III–V semiconductors. In melt growth experiments of III–V compounds, a floating zone has been used very seldom because of their too-low surface tension and a relatively high density of material at the melting point. It is therefore difficult to grow crystals from the melt with diameters greater than 6–8 mm[35, 36]. With decreasing growth temperature, the surface tension will increase, and the crucible-free growth of crystals with diameters greater than 10 mm should be possible. GaSb growth experiments have been done with a float solution zone of 10 mm diameter[26]. The growth temperature was 500 °C, and the growth rate 2,2 mm/day. In order to insure stable growth conditions over a period of days, the importance of a symmetric solution zone (zone length $\ell = 8$ mm) can be demonstrated (Fig. 12). The growth studies were carried out in a Pd-diffused

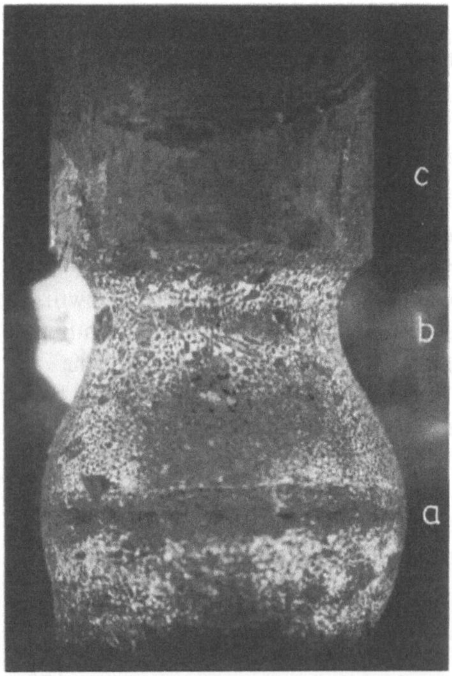

Fig. 12. Float solution zone of GaSb in a mirror Furnace[22]. $T_M \sim 500$ °C, *a:* grown GaSb crystal, *b:* Ga-Sb-solution, *c:* GaSb-feed; crystal diameter: 10 mm

hydrogen athmosphere with a heater of high-purity carbon. GaSb crystals up to 10 mm length were grown. The grown material was single crystalline with pronounced facets (3-fold symmetry due to the (111) seed plane).

THM-crystals have been studied through the use of various techniques, e. g., dislocation etching[24, 25, 37], electrical measurement[24, 38], and photoluminescence studies[26]. In TSM growth of GaAs[13, 20], GaSb[17] and SiC[39], the dislocation density in the grown crystal was found to decrease abruptly by an order of magnitude or more from that in the seed. Similar results have been reported for THM growth of GaAs[24] and GaSb[25]. Figure 13 shows the etch-pit density (EPD) as a

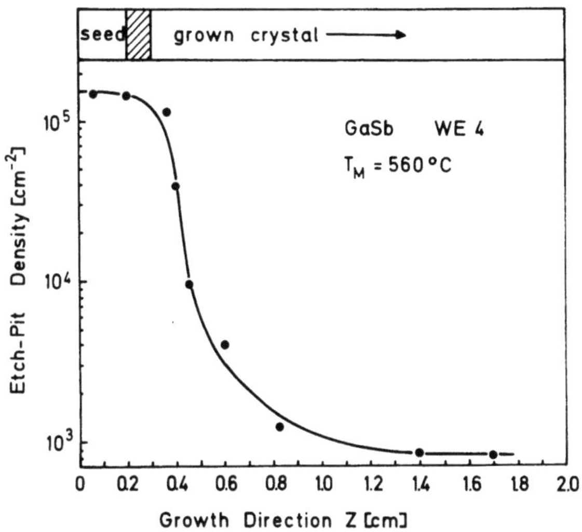

Fig. 13. Etch-pit density (EPD) against grown length of a GaSb sample (centre) EPD seed ≅ $2 \cdot 10^5$ cm^{-2}

function of the grown length of a THM-GaSb crystal (<111>)[25]. For dislocation etching the crystals were sliced perpendicular to the growth direction. The result shown in Fig. 13, which can be explained by several models (see below), is typical for crystals grown by THM in the <111> direction. The EPD decreases drastically within the first few millimeters of grown material and then reaches a saturation level. It should be pointed out that the dislocation density of GaSb and InSb is higher at the periphery of the crystals, which were grown in a quartz ampoule[25]. This behavior was not observed in THM growth of GaAs[24]. Horizontally grown as well a crucible-free crystals showed a homogeneous EPD distribution over the whole slice area down to about 100 cm^{-2}.

The following mechanisms have been proposed to explain the decrease of dislocation densities in the grown THM crystals:

(i) the sharp decrease of EPD within 1–2 mm after start of growth with <111> oriented seed crystals can be attributed to the mutual annihilation through dislocation due to the high dislocation density of the seed crystal. This annihilation mechanism is found either in pairs of dislocations of opposite Burger's vector annihilating one another by forming a loop, or by their transformation into other types of dislocations which grow alongside the surface[12, 37].

(ii) The lower decrease of EPD in the region above 2 mm from seed can be explained by an outgrowth via convex interface[37].

Figure 14 depicts the emission spectra of a typical GaSb seed crystal, a GaSb liquid-phase epitaxial layer (growth temperature 330 °C), and a slice of a GaSb-THM crystal grown at 420 °C (v = 1,5 mm/day[26]). The near-gap luminescence of the LPE

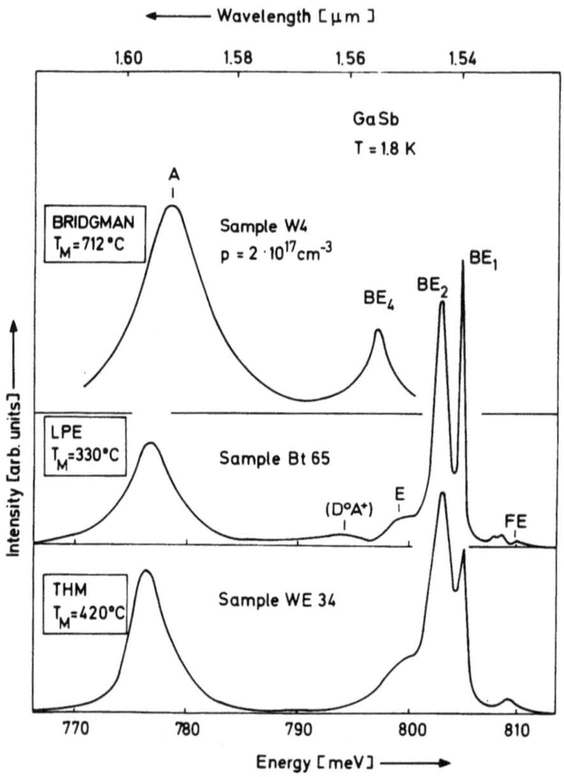

Fig. 14. Photoluminescence spectra at 1.8 K of GaSb grown by different methods.
Excitation ligth: Krypton-Laser λ = 514 nm, detection of luminescence: grid-monochromator with PbS cell

layer can be attributed mainly to the radiative decay of bound and free excitons[5]. The resolution of near-edge emission into separate lines and the half width of these lines can serve as a measure of the impurity concentration in the crystals. Line A in Fig. 14 is due to the native defect in GaSb, as demonstrated by Van der Meulen[3] [b]. By comparing the emission spectra of the LPE sample and the THM crystal (which are nearly identical), one can see the "LPE-quality" of the grown GaSb-THM crystal. In both cases the net hole concentration should be below $3 \cdot 10^{16}$ cm^{-3} [5], whereas the seed crystals has a hole concentration of $p \sim 2 \cdot 10^{17}$ cm^{-3} (typical value of melt grown GaSb crystals).

---

[b] The lines BE 1 and BE 2 represent the emission of an exciton, which is bound to the neutral acceptor A[40].

2.4 The Solute, Synthesis, Diffusion-Method (SSD)

The *S*olute, *S*ynthesis, *D*iffusion Method (SSD) has been developed by Kaneko et al.[41] in order to grow high-quality GaP crystals and to avoid the need for the expensive high-pressure growth apparatus used in the Czochralski liquid encapsulation technique for the growth of GaP[41]. The principal set-up of this method is schematically shown in Fig. 15. A crucible-containing liquid Ga is placed in a temperature

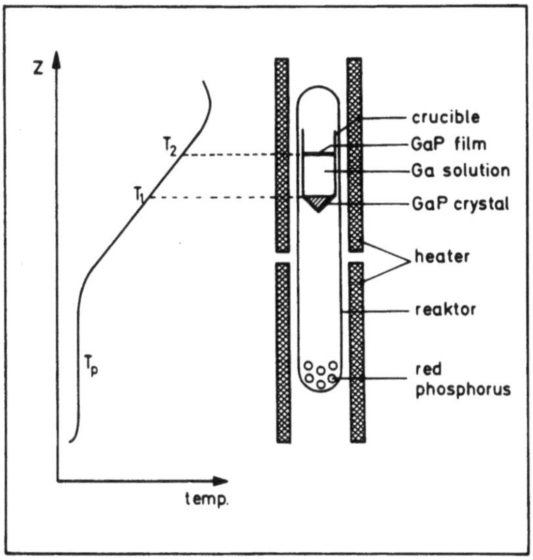

Fig. 15. *S*ynthesis, *S*olute, *D*iffusion method (SSD) (schematically) for GaP single crystal growth[41]

gradient and the upper surface is exposed to P vapour of about 1 bar. This means a phosphorous temperature of about $T_P \cong 420$ °C. GaP synthesized at the surface dissolves in the liquid Ga and the solute diffuses downward and saturates at the bottom; the crystal then starts growing from the bottom. To maintain the GaP film at the surface of the Ga melt, the phosphorous vapor pressure must be higher than the dissociation pressure of GaP at T_2. As in TSM growth (see Chap. 2.3.1), no moving parts are necessary in SSD growth. Various modifications of the SSD-method are possible[41, 44, 45]: (i) A charge travel equal to the growth rate. In this case, a constant temperature profile can be maintained at the growing interface (T_1 in Fig. 15); (ii) a seed crystal at the bottom of the crucible, and (iii) an open SSD system with a condensor at the top of the vessel to trap the phosphorous.

The pulling-down variant of the SSD method is similar to a method used for the growth of GaP crystals and reported by Rodot et al.[46] However, in contrast to SSD, an additional temperature gradient was applied to the growing interface.

The SSD method combined with a seeding technique has been mainly used for the growth of GaP crystals[41, 42–46]. A survey on various SSD-growth data of some III–V compounds is shown in Table 3. The growth rate is dependent on the temperature difference between the top of the solution (GaP surface) and the temperature of the growing interface and is limited by the diffusion process of the solute. The

Table 3. SSD-Growth parameters of various III-V compounds

III-V Semiconductor	Solvent Solvent	Solution Temperature T_2 °C	Temperature °C/cm	Growth rate mm/day	References
GaP	Ga	1065–1330	7–46	0,8–7	Keneko et al.[41]
	Ga	1100–1150	120	3,7–4,6	Gillessen and Marshall[43]
	Ga	~ 1200	15–30	1 –2,5	Moravec and Novotny[44]
$In_{1-x}Ga_xP$ $(0,1 \leqslant x \leqslant 0,8)$	In	800–1120		1,6	Marshall and Gillessen[52]
InP	In	850– 890		2 –3	Marshall and Gillessen[49]
	In	700– 830	20–39	0,16–0,43	Ito and Ito[50]

growth rate can be calculated in a similar way to Eq. (2) and fits very well the experimental values[41, 44]. GaP crystals of 19 mm diameter and with lengths up to 52 mm[43] have been replicated. The crystals were single crystalline with small grains near the crystal surface, which may have been caused by nucleation at the crucible walls[41, 43]. In order to control the conditions at the growth front during crystal growth, a weight-gain method and a phosphorous consumption method have been developed[45].

SSD-GaP crystals show low dislocation (D) pit[47] ($\leqslant 10^5$ cm^2) and very low saucer (S)[47] pit densities ($< 10^4$ cm^{-2}), when compared with LEC-GaP crystals. Undoped SSD-GaP-crystals were n-type with net carrier concentrations of about $1 \cdot 10^{15}$ cm^{-3} and room temperature mobilities of ~ 70 $cm^1 V^{-1} s^{-1}$. Main residual impurities are sulphur, carbon (graphite crucibles) and Si (quartz crucibles)[48]. SSD growth was also applied to InP. Single crystals were obtained up to 22 mm in diameter (maximum length about 42 mm) with flat-bottomed crucibles (quartz, vitreous carbon and BN) containing (111)B disc-shaped InP seed crystals[49]. The grown crystals were n-type with concentrations ranging from $5 \cdot 10^{16} - 10^{17} cm^{-3}$. EPD values of about 10^5 cm^{-2} were of the same order of magnitude as those obtained in SSD-grown GaP[43]. P-type doping of InP with Zn and Cd was made in a SSD-variant with presynthesized InP as a starting material[50]. Due to a low growth temperature of about 750 °C, Zn evaporation of the In-solution could be reduced.

The low growth temperature resulted in a very low growth rate of about 1–3 mm/week, due to the low value of the diffusion coefficient of P in In ($D_{750°C} \cong 5 \cdot 10^{-5}$ $cm^2 s^{-1}$)[51]. Hole concentrations for both Zn and Cd lay between 10^{17} and 10^{18} cm^{-3}.

Additional details concerning SSD-growth results and applications are given elsewhere in this volume[52].

3 Growth of Thin-Layer Single-Crystal Structures by Liquid Phase Epitaxy

3.1 Introduction

Most procedures employed to turn a single-crystal semiconductor into an electronic device rely on processes applied to or acting through the surface: oxidation, etching, doping (e. g., by diffusion, ion implantation), metallizing, contacting (bonding) etc. Hence, very thin functional layers on a supporting substrate are adequate for most technical applications. In terms of purity and structural perfection, excellent quality for these layers is needed. Layers grown by liquid phase epitaxy (LPE)[53] meet these requirements in numerous optical and electronic devices; LPE techniques adapted to the demands of device technology are steadily being improved. Comprehensive reviews on various aspects of LPE have already been presented by Dawson[54], Kressel and Nelson[55], Giess and Ghez[56], Deitch[57], Brice[58], Randles[59], Zschauer[60], and Foster[61]. In this chapter, therefore, we shall place the emphasis upon some of the parameters influencing growth and structural properties of LPE layers which have not yet been discussed in detail in other reviews. Although our discussion is restricted to binary III–V compounds, it occasionally deals with other compound semiconductors and garnets as well.

3.2 Basis of LPE Growth Techniques

In principle, liquid phase epitaxy is a simple and, consequently, a reliable process. It is comprised of

i exposing a substrate wafer to a solution whose composition is that of the liquidus of the binary system (or of a ternary alloy system[61]).

ii cooling the saturated liquid in a strictly controlled manner in order to grow the desired film.

iii removing the supernatant liquid in order to terminate layer growth.

Growth parameters can be easily controlled and layer properties can therefore be readily reproduced. Careful attention, however, must be focused on all three phases of LPE growth to insure that suitably doped layers of proper thickness, proper flatness, and low defect density can be grown.

3.2.1 Choice of Solvent

The solvent has a strong influence on the properties of the epitaxial layer. It is generally a melt of one of the group III elements, when layers of the III–V–compounds are prepared. In any case, the solvent must dissolve sufficient quantities of the solute in order to obtain layers of usable thickness. Growth of high-purity layers, in particular, requires solvents which are themselves components of the solute. If other low-melting solvents are used, such as Sn, Pb, Zn, Bi etc. (for solvents, see Ref. 57 and references therein), atoms of the solvent are incorporated to a certain extent into the growing crystal. Thus, the resulting semiconductor is doped with an electrically or optically active element.

3.2.2 Growth Systems

Many of the growth systems used today were developed as a result of numerous experiments conducted with other designs. LPE is practiced in three major ways: tipping, dipping, and sliding. The first growth systems explored were the tipping device[53] in 1963 and the dipping device[62] in 1967. Both systems have been found appropriate for growing single, nominally homogeneous layers of most of the III–V compounds. Their application is obvious and straightforward (Figs. 16a, b and 17) and is described in detail elsewhere[53–63] and references therein. The original tipping furnace as well as the dipping apparatus have since been perfected into systems which permit simultaneous growth on many large-area substrates[64–66]. An important step in LPE technology was the introduction of multicompartment slider systems[67, 68], which have been employed for the growth of multilayered structures (Fig. 16c). These systems, and modified versions thereof, such as rotating drum

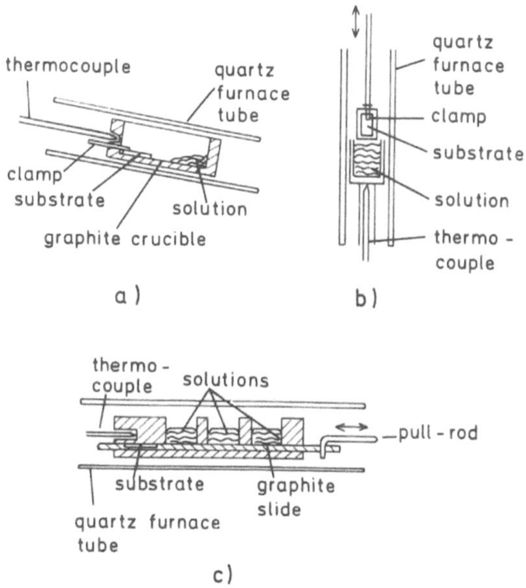

Fig. 16 a–c. Schematics showing the basic liquid phase epitaxial growth techniques.
a *Tipping* furnace as developed by Nelson[53)]
b Vertical *dipping* device after Rupprecht et al.[62)]
c *Multicompartment slider* boat after[67)] and[68)] for the sequential deposition of several layers during one growth cycle

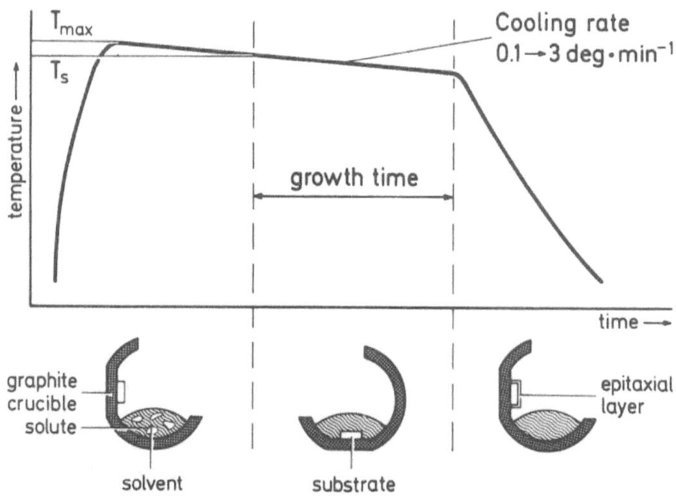

Fig. 17a. Temperature during one growth cycle in a tipping furnace. The substrate is exposed to the solution at temperature T_S. Schematic of modified "tipping" crucible after Grobe and Salow[63)] and Bauser et al.[89)]

Fig. 17b. Graphite tipping crubicle of Fig. 17a, supported by elastic quartz rods, after Bauser et al.[89]. The crubicle is shown in the position when layer growth was terminated. A small droplet of solution residue is visible at the lower edge of the substrate

devices[69, 70], were developed mainly to prepare GaAs-GaAlAs heterostructures. These consist of two or more successive layers of two semiconductors. Heterostructures are required for such devices as mode confinement lasers[71] and solar cells[72]. Multicompartment slider systems are usually automated for production purposes[73]. When the demand for multilayered structures of semiconductors arose, tilting- and tipping-type growth systems were designed to allow multiple-layer growth in one epitaxial run[74, 75].

Descriptions have frequently been published of epitaxial systems suited to the growth of layers satisfying particular requirements. For instance, very uniform, thin

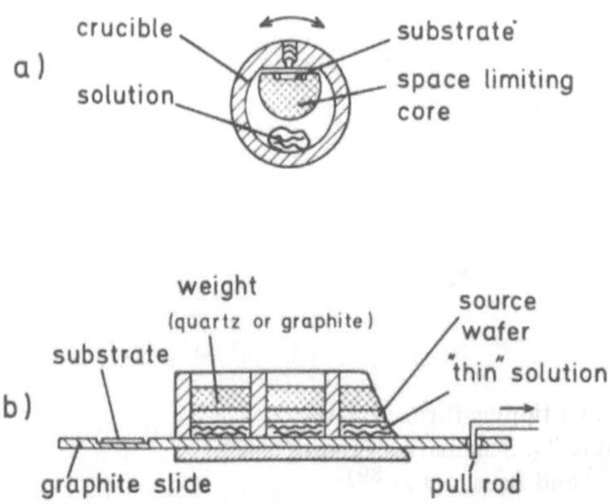

Fig. 18 a and b. Schematics of boats providing for the growth of very uniform thin layers. **a** multicompartement slider boat after Lockwood and Ettenberg[77]. In both cases small solutions are flattened against the substrates: (**a**) by a space limiting core and (**b**) by a quartz is often referred to as "thin solution epitaxy"

layers are obtained when the growth solution is forced to spread through a narrow gap between the substrate and a space-limiting plate; this growth condition has been realized in tipping and dipping systems[76, 66] and in slider devices (Fig. 18)[77]. A well-defined temperature gradient in the direction perpendicular to the substrate face was achieved by applying auxiliary heaters[78]. Such temperature gradients have been reported to reduce terracing of the layer surfaces (see Chap. 3.3). Isothermal conditions for growth in different compartments are achieved in a vertical rotating drum with symmetry around the vertical axis[69]. Sliding or moving parts, like those used in normal multicompartment devices, are avoided in the wipeless boat design[79]. The depositing of extremely thin LPE layers from a fast-flowing solution was accomplished in a device which utilizes centrifugal force[80].

3.2.3 Temperature Range and Growth Rate

The growth temperatures in LPE range from about 350 °C to about 900 °C. In many cases, temperatures at which LPE layers are deposited are lower than corresponding temperatures in VPE processes. LPE growth rates are nevertheless high enough for efficient layer production. Growth rates between 0.001 and 10 $\mu m \cdot min^{-1}$ are possible, with usual rates ranging between 0.1 and 3 $\mu m \cdot min^{-1}$. A low temperature, combined with a moderately high growth rate, reduces the danger of contamination in the growing layer. III–V Compounds, when grown by LPE, usually have fewer unwanted recombination centers for a given doping level than are present in GaAs when it has been prepared by other growth techniques[81]. The density of point defects, such as traps and nonradiative centers in luminescence, has been reported to be lowest in LPE material[82]. This explains the high efficiency of many LPE-grown optoelectronic devices[81]. High-purity LPE layers of GaAs and InP have also proved successful in their application to microwave devices[60]. Low growth temperature and high growth rate make LPE an economical means for the preparation of high-quality layers. These findings justify further efforts to develop LPE techniques from small-scale laboratory applications to large-scale commercial semiconductor growth operations.

3.3 Growth Mechanisms and Surface Morphology

3.3.1 Influence of Substrate Morphology in Growth Mechanism

LPE layers grow according to a "lateral" growth process, as is typical for crystal growth from solution within the usual temperature range (c. f. Sect. 3.2.3)[83, 84]. New molecules are incorporated into the crystal at the edges of steps (Fig. 19), and steps are practically always present on the growth interface. Consequently, as growth proceeds, the steps move in a "lateral" direction approximately parallel to the growth plane. This general process comprises several possible growth mechanisms, which of these growth mechanisms will dominate depends on the geometrical configuration of the steps initially present at the substrate surface. Hence, one of the most important parameters in LPE growth in the initial substrate morphology which

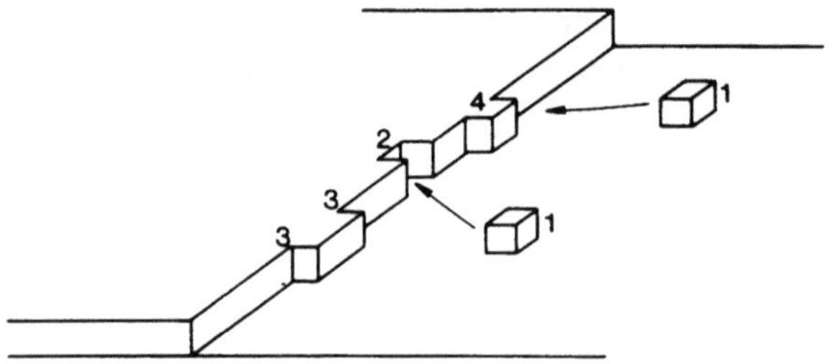

Fig. 19. Edge of a surface step with typical defects. Adsorbed surface molecules (*1*) become incorporated into the crystal in edge vacancies (*2*), at kinks (*3*), or at straight portions of the step as adsorbed edge molecules (*4*)

exists at the point where epitaxial deposition sets in. In the following section, the influence of substrate morphology and of growth conditions on homoepitaxy is discussed. The effects described apply also to heteroepitaxy, although in this case, additional complications arise due to lattice mismatch.

3.3.1.1 Orientation Effects

In the following discussion the substrates are assumed to have been prepared in the conventional way; i. e., their surfaces are planar and have a shiny, smooth appearance (on a macroscopic scale). The isotherm planes of the growth system are assumed to be parallel to the growth face of the substrate. Following these assumptions, the orientation of the growth faces of the substrates determines the particular growth mechanism. Let us first consider the most common case encountered in practical LPE work. Usually, the orientations of the substrate growth faces are "nominally" (100) or (111). The accuracy of these nominal orientations is often not better than 0.5 deg. to 1.5 deg.[c]. Consequently, before growth starts, the substrate surface itself will exhibit misorientation steps. These steps extend along the intersection lines of the substrate surface and the (100)- and (111)-crystal planes respectively. Misorientation steps of this type have been detected, for instance, on Silicon substrates[85]. Their heights and mutual distances are randomly distributed. Under such a particular starting growth condition, "transient growth"[86] will occur. During the "transient" period, both the profile of the growth interface and the rate of growth R (Fig. 21) may change. A step-bunching process occurs during the formation of the layer[87]. The step-bunching causes an increase in the average height h and distance λ of steps at the growth interface (Fig. 21). Due to this increase in height and distance, the steps become easily visible on the surface of thick LPE layers and are usually referred to as surface terraces[88–90] or rippled structures[91–96]. Surface terraces are one of the most common defects in LPE growth (Fig. 20).

---

[c] (for the angle of misorientation θ see Fig. 21)

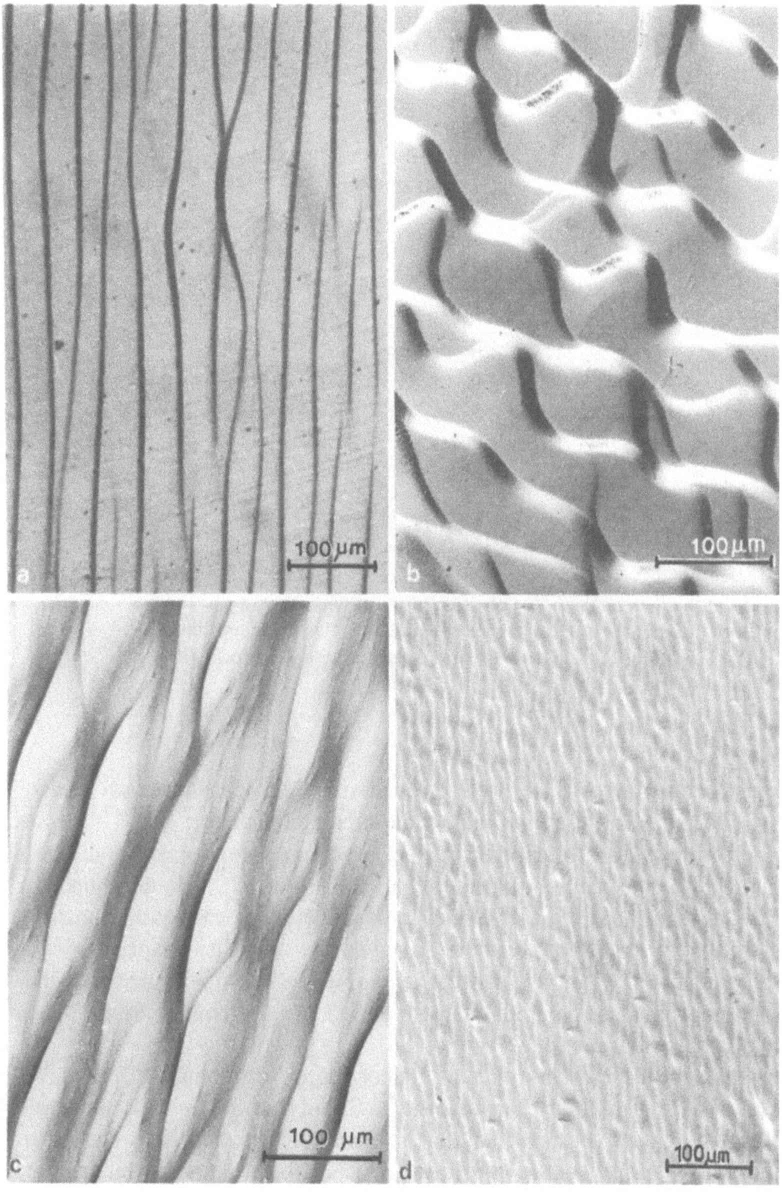

Fig. 20 a–d. Surface terraces of nominally (100)-oriented, undoped GaAs layers, **(a)** grown on a smooth substrate. The substrate was misoriented by an angle θ = 28′ from a (100)-crystal plane. T_S = 745 °C, layer thickness 24 μm. **(b)** grown on a rough substrate, T_S = 720 °C, layer thickness 16 μm. **(c)** grown on a smooth substrate, T_S = 788 °C, layer thickness 110 μm. **(d)** grown on a smooth substrate, T_S = 460 °C, layer thickness 1 μm

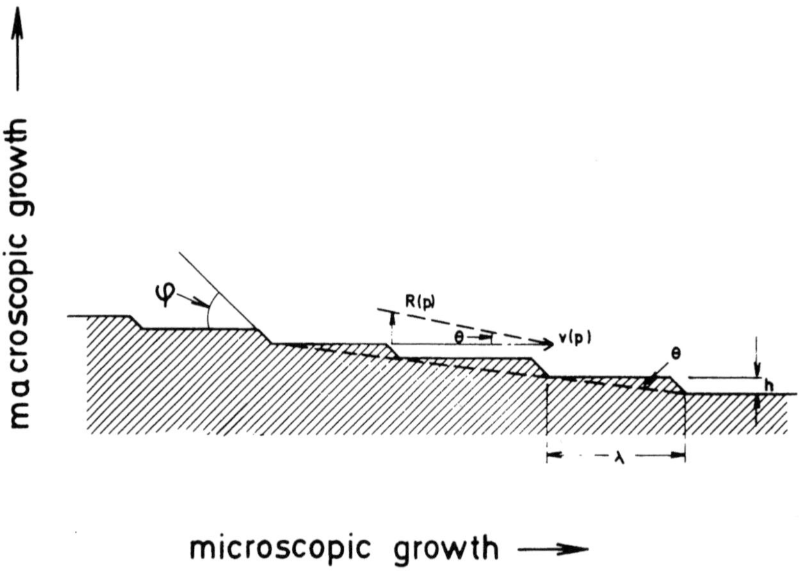

Fig. 21. Profile of a growing crystal surface of orientation p, with p = tang θ. The growth face is misoriented by the angle θ from a low index lattice plane. ν is the horizontal velocity of the steps of height h. The interstep distance λ is h/p. R is the rate of growth normal to the orientation p = o (after[86])

In earlier reports, surface terraces were often attributed to constitutional supercooling, in terms of the growth rate normal to the substrate plane[93, 94, 97, 98]. More recent work, however, demonstrates that this special surface morphology should be reinterpreted in view of the accuracy of recent substrate data[88–92]. In addition, Peters[88] showed for GaP that the growth can be described as a near-equilibrium process, with the thickness of the layer determined by the solubility change of phosphorus in the solution within the range of an appropriate temperature interval. The growth efficiency in these experiments was found to be determined by the geometry of the epitaxial system.

Since surface terraces are undesirable in semiconductor technology, efforts have been made to eliminate them. Mattes and Route[92] found that for GaAs, a temperature gradient normal to the substrate reduced the terrace width and height due to an increased "nucleation density" at the substrate surface. Other techniques reported by Hsieh[100] to be successful in restraining the development of terraces are step-cooling[100], supercooling[101], and transient-mode liquid epitaxy (TMLE)[102, 103]. In the supercooling technique the substrate and growth solution at a uniform rate are cooled to a temperature below the liquidus temperature (but not low enough for spontaneous precipitation of GaAs to occur in the supercooled solution): both are then brought into contact and cooled without interruption at the same rate until the solution is wiped off the substrate, thereby terminating growth. In the step-cooling procedure[100], the substrate and solution are cooled to a fixed temperature below the liquids temperature (again, not low enough for spontaneous precipitation). They are brought into contact at this temperature and kept at the same temperature until growth is terminated. The supercooling and step-cooling techniques have been reported to yield very smooth, flat layers of reproducible thickness in a range of from one-tenth to tenths of micrometers[100)]. The smoothing effect in the supercooling and step-cooling techniques may be ascribed to an increased density of nucleation sites in the initial stages of epitaxial deposition. An increased density of nucleation sites tends to reduce the average height and distance of the terraces, thus favoring the development of smoother surfaces.

*T*ransient-*M*ode *L*iquid *E*pitaxy is another way of inducing fine nucleation by supercooling. TMLE is accomplished by bringing a cooler substrate into contact with a hotter saturated or supersaturated growth solution. Immediate cooling of the solution at the substrate surface causes a high degree of supersaturation. As this supersaturation disappears, an epitaxial layer grows. The growth rate increases rapidly at first and then decreases as the solution temperature adjacent to the substrate surface approaches the average temperature of the solution. TMLE techniques have been extended to the growth of epitaxial layers with a large degree of lattice mismatch. The superior quality of these layers has been compared to those grown by equilibrium-ramp cooling and supercooling techniques[103)].

Another successful way to eliminate terracing in LPE is through accurate substrate orientation. The tendency toward terrace formation may be reduced considerably by either increasing the substrate misorientation to more than 2 degrees[91)], or by decreasing the substrate misorientation to below 0.1 deg.[88, 89, 90)].

a) Increasing the Substrate Misorientation, Critical Orientation. According to a theory formulated by Rode[91)], a "critical" orientation of the substrate exists, preventing the formation of terraces during epitaxial growth. The critical orientation corresponds to a misorientation angle θ, which equals the step – to – riser angle on a terraced layer (Fig. 21) grown under otherwise equal conditions. Experimentally, the exact value of the critical angle is determined by taking a surface profile of the terraced layer and by adjusting the misorientation of the layer accordingly.
For the growth of InP on (100)- and (111)-oriented substrates, Messham and Majerfeld[104)] reported the "critical" misorientation angle θ to be 2.6 deg. and 3.3. deg. respectively. In the case of GaAs, an example of the angle θ has been given 0.8 deg.[105)]. The misorientation is in all cases towards the <110> direction. The step-to-riser angle and, thus, the critical misorientation usually increase with greater layer thickness. Misorientation angles are expected to cover the range of up to several degrees for layers with thickness exceeding 50 μm. According to Rode's theory[91)], the surfaces of thick layers, grown on critically oriented substrates, exhibit a "long-wavelength" sinusoidal structure (Fig. 22).
b) Accurate Orientation, $\theta < 0,1$ deg. (1) Elimination of surface terraces was obtained by Peters [88)], Bauser[89)], and others, when the substrate misorientation was reduced to below 0.1 deg. (Fig. 23a). The surfaces of layers grown on such accurately oriented substrates are smooth and exhibit only growth steps of extremely small height. Surface steps of such small height are practically invisible in the microscope. An epitaxial "decoration" technique was therefore applied to render these steps visible.[106)]. The effect of decoration is schematically illustrated in Fig. 23b. A typical example for the arrangement of steps of small height, which have been

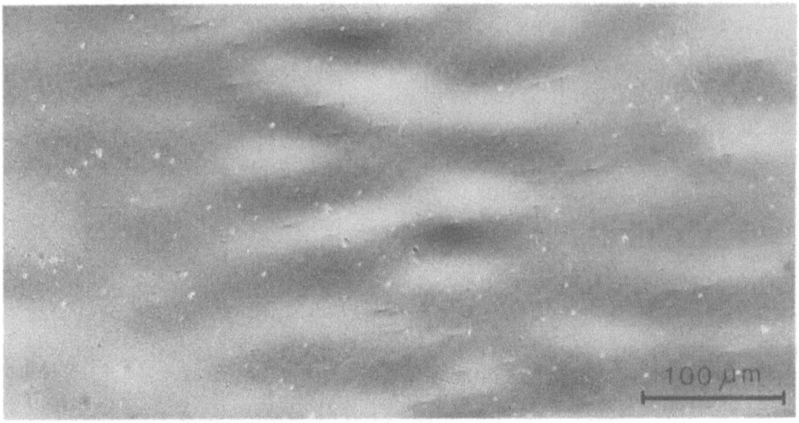

Fig. 22. Surface of a 50 μm thick GaAs LPE layer, grown on a substrate with θ = 3 deg. from the (100) plane

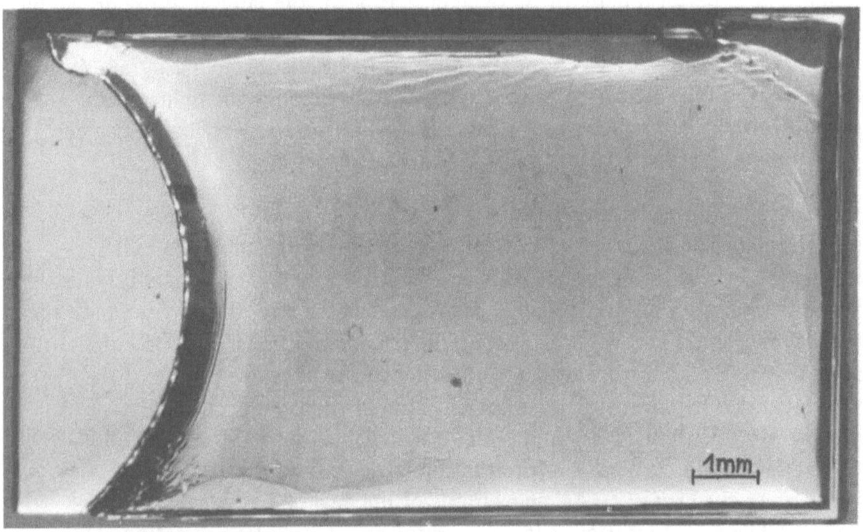

Fig. 23a. Epitaxial layer grown on a carefully oriented substrate. The misorientation θ from the (100) plane is less than 5'. Terraces did develop only near the edges, where the substrate surface was misoriented because of bevelling by the polishing and etching processes

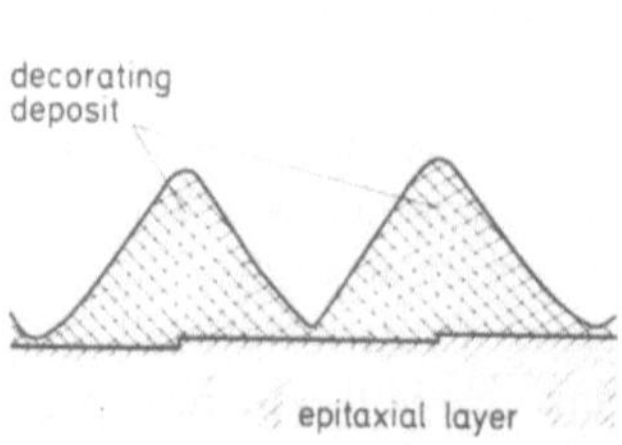

Fig. 23b. Profile of a crystal surface with growth steps of small height being decorated (schematic). Decoration is achieved by selectively depositing an appropriate substance along the edges of the growth steps (according to Bauser et al.[106])

revealed by decoration, is shown in Fig. 23c. No tendency to develop terraces has been observed for such layers, not even in the case of very thick layers.

Fig. 23c. Decorated growth steps on the surface of a smooth GaAs LPE layer[106]. The layer is approximately 50 µm thick. The growth steps are arranged rather regularly yet not strictly as regular as in the case of Figs. 25 and 26

(2) Faceting. When the accuracy of the substrate orientation is further improved, facet growth may occur. Facets have often been observed along an edge or in a corner of epitaxial layers grown on slightly misoriented substrates (Fig. 24,a,b,c).

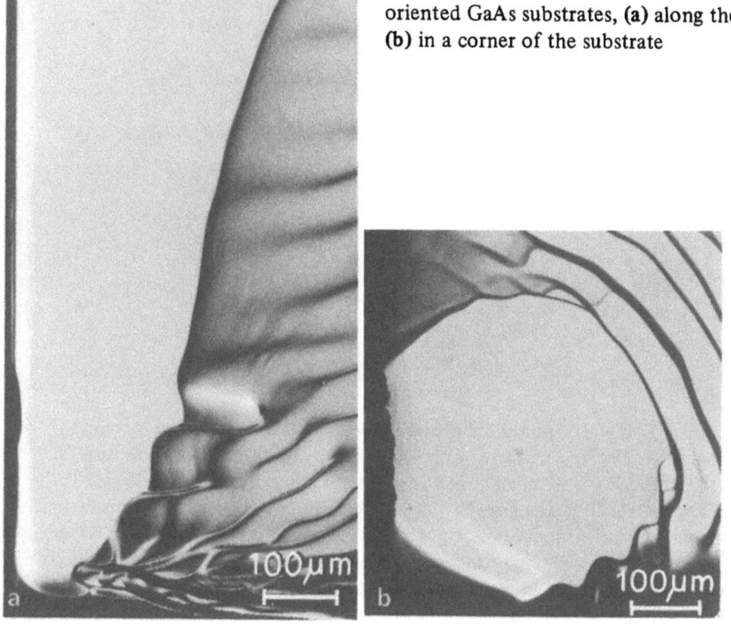

Fig. 24a and b. Facets which developed on GaAs LPE layers during growth on nominally (100)-oriented GaAs substrates, **(a)** along the edge and **(b)** in a corner of the substrate

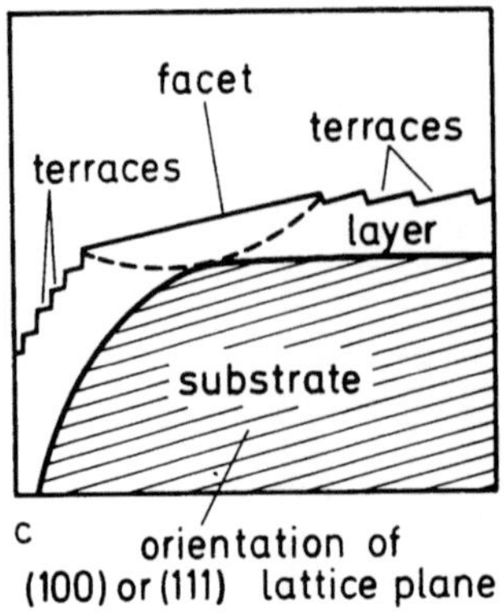

Fig. 24c. Schematic cross section of an epitaxial layer on a misoriented substrate near the edge of the layer. Facet formation in GaAs usually begins, where the curvature of the growth interface passes through a (100) or a (111) plane

Fig. 25. Growth step pattern on a (100)-oriented growth facet of a GaAs LPE layer (interference contrast micrograph). The step pattern was revealed by decoration. The decoration material being deposited in the shape of small cones or pyramids along the steps (see Fig. 23b) may entirely coat the original surface. The growth steps are clearly polygonized and thus form four-sided growth pyramids. Spiral step patterns as well as step patterns with concentric closed loops were observed on (100) facets. A few surface areas which are atomically flat, are visible in the left as well as in the right half of the micrograph. These areas are not covered by the decoration deposit. The steps are of monomolecular height[108])

The condition favoring facet formation arises when the solid-liquid interface is parallel to a (100) or a (111) plane – at least locally[107]. Since facets represent slowly-growing faces, they usually spread as epitaxial growth proceeds[83]. By taking advantage of this feature and by exactly orienting the substrates parallel to (100) or (111) in the first place, facets may be grown in such a manner that they extend over large areas. The misorientation must be below 1 min. of arc.

Facet growth results in macroscopically planar and mirror smooth growth interfaces and layer surfaces. The planar surface structure of factes of GaAs LPE layers has recently been shown to result from a strictly periodic arrangement of growth steps[108]. The steps are usually of monomolecular height. Figures 25 and 26 show

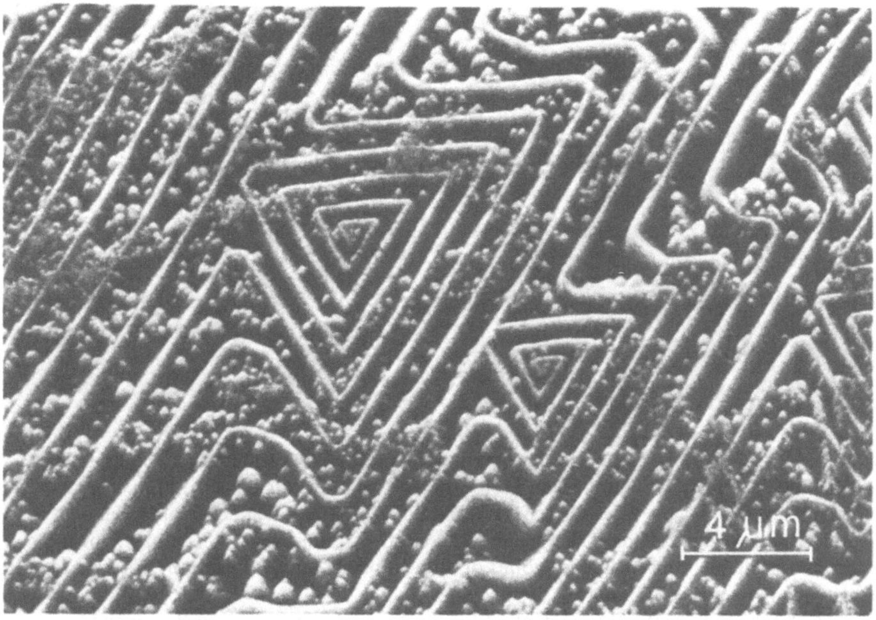

Fig. 26. Typical step pattern on a (111) facet of LPE GaAs (scanning electron micrograph). The individual steps appear as a string of close adjacent cones or pyramids. The growth steps are polygonized and clearly reflect the three-fold symmetry in the (111) crystal plane[108]

typical growth-step patterns on (100)- and (111)-oriented growth facets. The monomolecular growth steps were rendered visible by decoration[106]. Upon closer examination, the steps form spirals or even closed loops and thereby shape the surface into a pattern of very low-profile cones or pyramids (Fig. 27a).

According to a suggestion made by Frank[109, 110], these steps are provided by dislocations which terminate in the growth interface, i. e., the crystal surface, and have a screw component perpendicular to the growth interface. The operation of step sources of this kind is described in detail by the "Frank screw dislocation mechanism"[109, 110]. In this model, a single dislocation with a screw component perpendicular to the growth interface creates a growth step spiral. Closed loops of steps

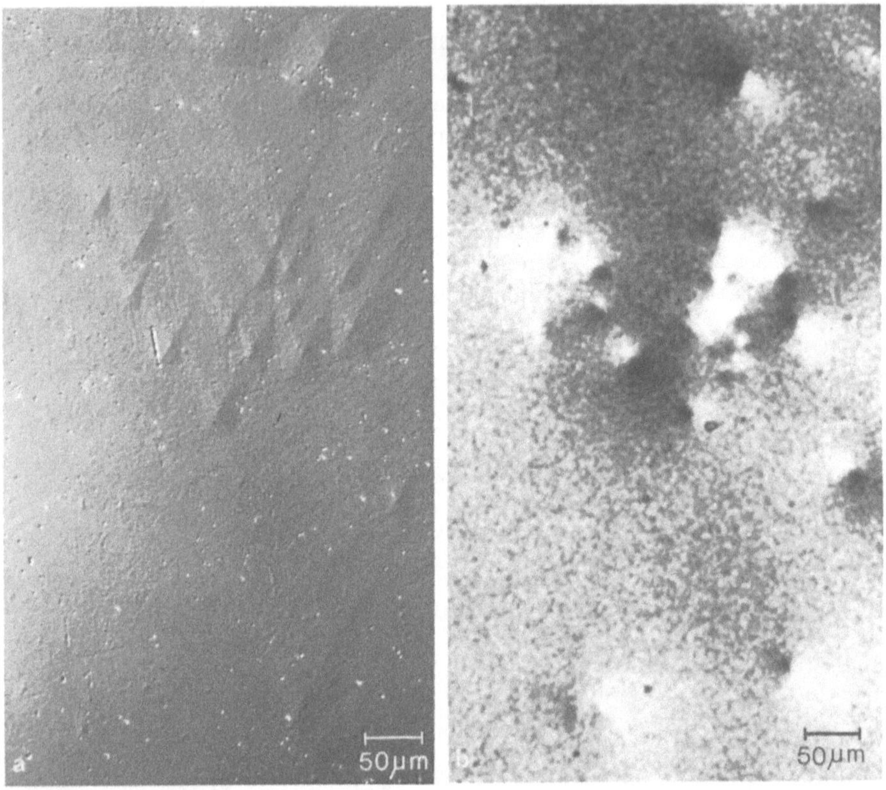

Fig. 27. a Surface area of a GaAs LPE layer exhibiting facet growth (interference contrast micrograph). The substrate orientation is (100), the average layer thickness is 18 μm. The micrograph reveals, that the facet surface is not flat on an atomic scale, but is composed of low-profile, foursided pyramids. The pyramids are set off by faint shadings. The apexes of the pyramids coincide with the emergence points of dislocations in the topograph (**b**). The average height of the pyramids is < 30 nm.

b X-ray, double-crystal topograph of the facet area shown in a) after Bauser and Hagen[111]. A distinctive contrast pattern marks the emergence points of dislocations in the layer surface. There is a perfect one-to-one relationship between the apexes of the pyramids in (a) and the emergence points of dislocations in (**b**)

emerge from a pair of dislocations with opposite screw components[110]. Cones or pyramids develop if the rate of advance of the steps is isotropic or anisotropic respectively (see Figs. 25 and 26). This model was recently verified in experiments for GaAs LPE-grown facets by Bauser and Hagen[111]. They applied X-ray topography in combination with interference contrast microscopy (Fig. 27) and the decoration technique reported in[106]. In addition Bauser and Strunk[112] recently reported the first microscopic proof of the existence of the Frank screw-dislocation mechanism. Applying high-voltage transmission electron microscopy, they showed the exact coincidence of the emergence points of dislocations with the apexes of growth-step pyramids for LPE-grown GaAs. They showed, in addition, that even dislocations without a Burgers-vector component perpendicular to the growth interface act as step sources. The efficiency of these "edge dislocation step sources" has been reported to be similar to the efficiency of Frank's screw-dislocation step sources[112].

We felt it necessary to dwell upon growth mechanisms in this section, because these may be taken into account in the further development of LPE layer growth in modern material technology.

3.3.1.2 Surface Oxidation Effects

Irregularities at the substrate surface may strongly influence the morphology of the epitaxial layer. If the substrate surface is partly oxidized, or marked by other impurities, 'nucleation' on the substrate crystal for the epitaxial growth is impaired, and island growth may be induced[88]. An example of island growth on a partly oxidized substrate is shown in Fig. 28. The substrate is misoriented from the (100) plane by

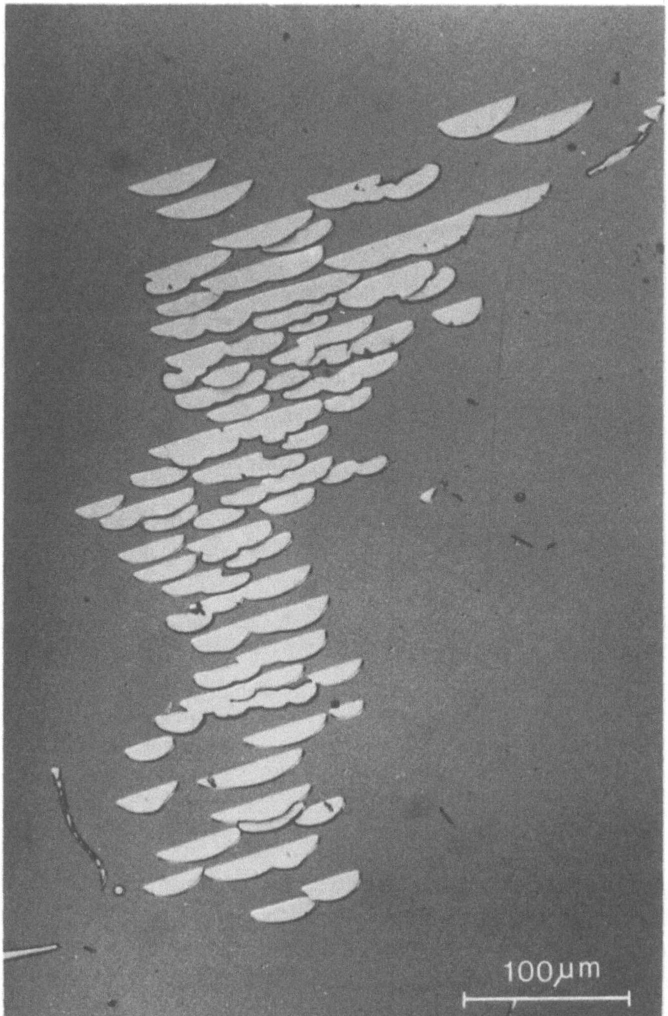

Fig. 28. Island "nucleation" on an oxidized GaAs substrate, which is slightly misoriented from the (100) plane. The nucleation occurs along intersection lines of the substrate surface and (100) crystal planes, i. e., along the edges of misorientation steps at the substrate surface. Epitaxial growth was started at $T_s = 450\,°C$. Nucleation phenomena on substrates of different orientations were described by Mattes and Route[92]

approximately 0,8 deg. The initial epitaxial growth (at 450 °C) set in along the edges of misorientation steps. This surface structure may illustrate how a low density of nucleation sites enhances the formation of terraces. The micrograph of Fig. 29 illus-

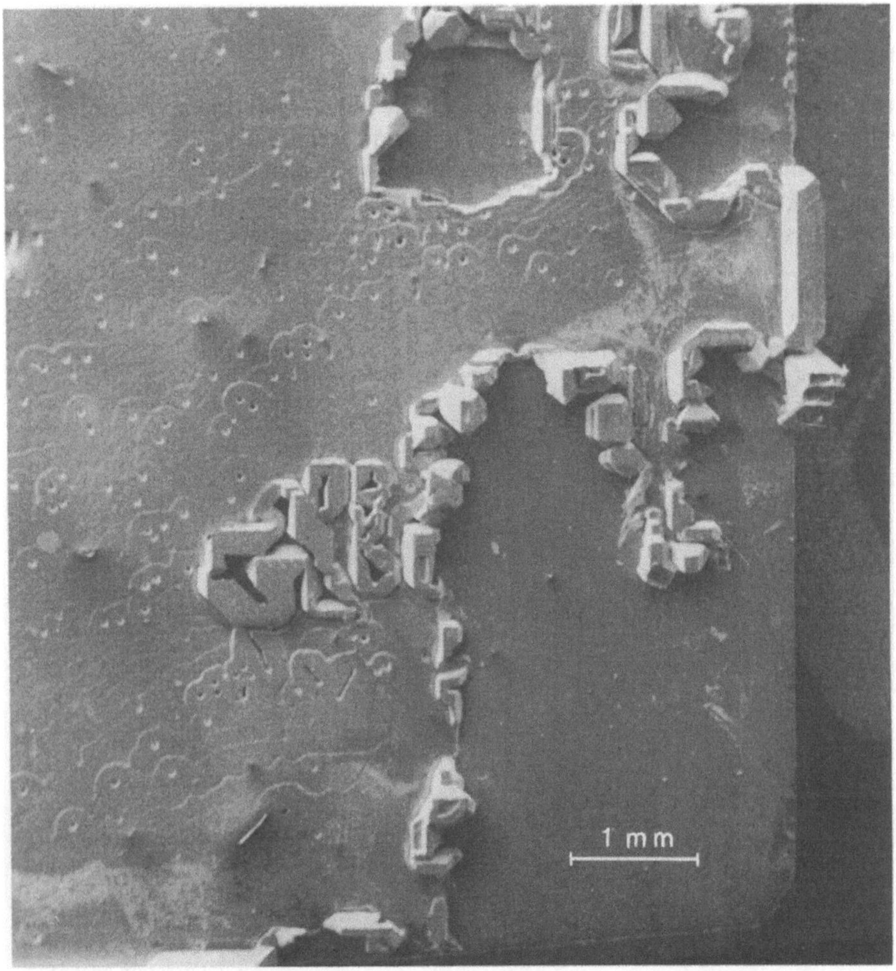

Fig. 29. Substrate which was held several times by a pair of tweezers. The tweezers were insufficiently cleaned and left enough of unspecified impurities on the substrate surface to inhibit epitaxial growth (scanning electron micrograph). The epitaxial layer surrounding this area is marred by numerous depressions and voids. These are caused by small splinters of GaAs resulting from handling the substrate with the steel-tweezers

trates the drastic masking effect of unspecified impurities. Several times this substrate was held by a pair of tweezers which were insufficiently cleaned beforehand, and traces of the tweezers show up on several spots of the epitaxial layer. These are but two examples of the way impurities at the substrate surface can induce growth irregularities.

3.3.1.3 Surface Asperities

All deviations from a smooth substrate surface tend to mar the epitaxial layer. Surface roughness should be avoided on the substrate from the very beginning since surface irregularities may be strongly enhanced during growth, in particular when a slight misorientation exists (see, for instance, Fig. 33). This holds especially true for the growth of multilayer structures, e. g., for layers of GaAs and GaAlAs, since irregularities often increase in size during a multilayer deposition.
A polished and accurately oriented substrate surface is often not sufficient to obtain smooth LPE layers. If the substrate crystal has striations, these can be altered to form surface asperities by etching the crystal. An example is reported for GaSb in which the etching worked selectively, thus slightly deepening the substrate surface along the striations[113]. The local misorientation caused by this slight depression may be sufficient to initiate terrace development. In the example given in Fig. 30, substrate misorientation and striations both gave rise to terraces which developed simultaneously, though in different orientations.

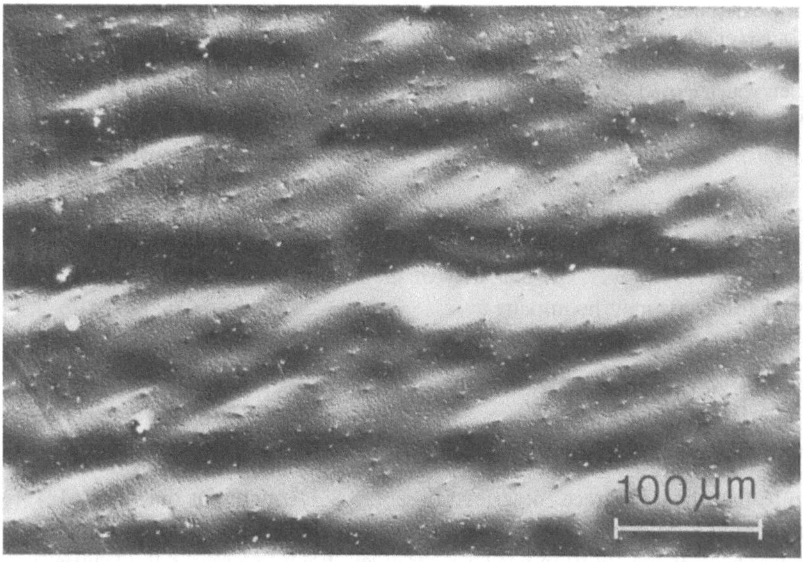

Fig. 30. Surface of a GaSb LPE layer with terraces caused by striations in the substrate crystal (*horizontal structure*) and by substrate misorientation (*diagonal structure*). After Linnebach et al.[113]

Another possible cause for a formation of surface asperities is thermal decomposition at the substrate surface during the heating which takes place prior to epitaxial deposition (Fig. 31). This applies especially to substrates of InP, which readily decompose because of the high vapor pressure of phosphorus, and which, therefore, must be protected against a loss of this component. Whether or not pits caused by

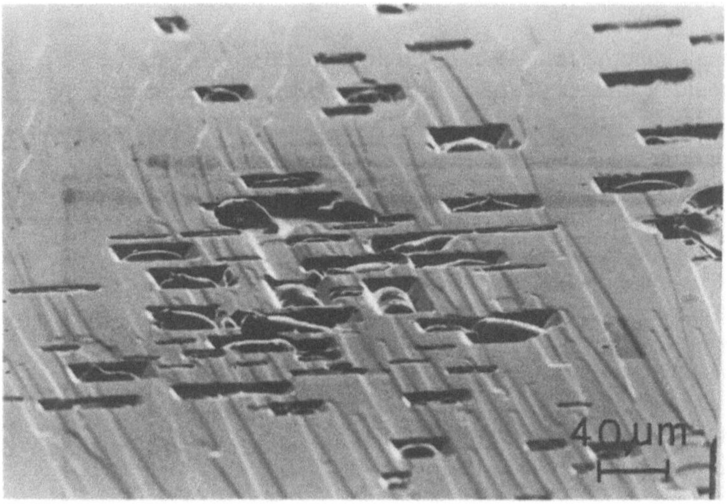

Fig. 31. Thermal decomposition at a (100) oriented GaAs substrate. The substrate was heated for 1 h at 850 °C in dry Hydrogen. The rectangular etch-pits are filled with Ga droplets. Shallow grooves ending in the pits are the traces of the Ga droplets which moved over the substrate surface (usually in 110 direction), in accordance with Kuphal[114]

this type of thermal etching severely corrugate the epitaxial layer depends on their size and depth. Occasionally, the pits are completely refilled and the growth face becomes smoother[114]. However, the planarity of the substrate-layer interface and the uniformity of thickness may be impaired.

3.3.2 Influence of Growth Conditions

3.3.2.1 Meniscus Lines

In addition to the surface terraces described in sect. 3.3.1.1 meniscus lines (Fig. 32) are another common defect. These are line-shaped defects which appear on the surface of the layers after the solution has been removed. Each line marks the transient position of a three-phase boundary line at the front or the tail of the solution during advancing or receding, respectively, on the solid surface[115, 116]. Apparently, the solutions move with a "stick-slip" motion; these features develop in places where the moving meniscus momentarily pauses. As already indicated, two sets of meniscus lines usually develop during the growth of a single LPE layer: one set, when the solution is applied to the substrate, and another set, when the solution is removed after growing the layer. The cross section of the lines has an approximately S-shaped contour with a surface depression pointing in the direction in which the gas-liquid boundary has moved[115]. Meniscus lines spoil the planarity of the growth interfaces, especially in multiple-layer structures. Several methods have therefore been proposed to eliminate them: the wipeless boat[79], the carry-over of a thin solution film[117], and a quick supply and removal of the solution[80]. The elimination of meniscus lines is especially important when extremely thin epitaxial layers are grown.

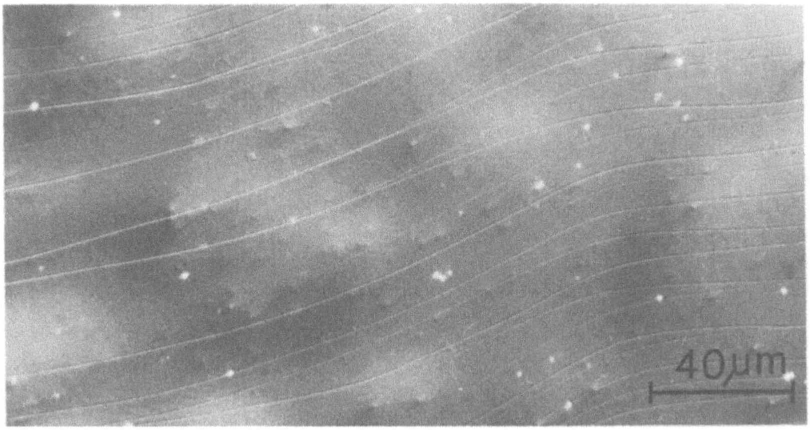

Fig. 32. Meniscus lines on the surface of a GaSb layer grown in a multicompartment slider boat. The direction of slider motion is indicated by arrows[113]

3.3.2.2 Microparticles and Microcrystals

Other defects, often encountered in LPE layers, are small depressions and voids[116]. These defects show up distinctly in layers of about 20–80 µm thickness (Fig. 33a). The depressions extend from the surface of the layer into the epitaxial material. Generally, a small "void", or hole, is found underneath each depression at the substrate-layer interface (Fig. 33b). Causes for the development of these defects may be small particles which are abraded from crucibles by sliding parts. Another source of small particles is the presence of microcrystals of the solute, which frequently are nucleated and develop in supercooled solutions[118]. Other possibilities for the development of depressions and voids are non-planar or rough substrate surface[118, 119], and an improper wetting of the substrate surface, resulting from oxides or other masking substances (c. f. Sect. 3.3.1.3). These defects can be avoided in epitaxial experiments.

3.3.3 Striations and Terrace Growth Model

In homogeneities in dopant distributions, usually called dopant or impurity striations, are a well-known phenomenon in bulk-grown semiconductor materials, in liquid-encapsulated Czochralski (LEC) growth of GaP[120], for instance, or in the Czochralski growth of InSb, Ge and Si[121, 122]. Recently, striations have been reported in LPE layers of GaP and GaAs, too[123–125]. The striations could be revealed on (110) cleavage planes perpendicular to the (100) growth plane of the layers. For both materials, two different types of striations were detected. "Type A striations" (named after Faust et al.)[126] usually consist of a family of continuous parallel lines which extend throughout the whole layer, regardless of the particular growth mecha-

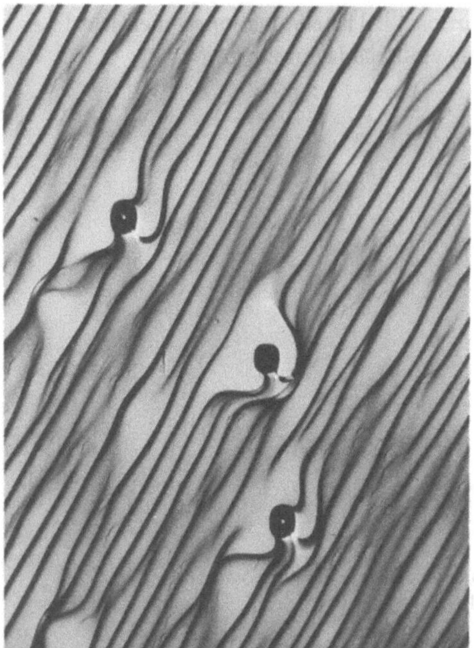

Fig. 33 a and b. Depressions in a terraced region of a GaAs layer[118]. **a** Plan view of the depressions in an early stage of development [layer thickness roughly 45 µm, substrate orientation nominal (100)]. The direction of step motion is from the upper left to the lower right corner of the micrograph. In a later stage of development, the depressions are surrounded by approximately semicircular terraces. Interference contrast. **b** (110) cleavage plane through an 80 µm thick layer with closely adjacent depressions, showing the presence of a void under each depression. The flat bottoms of the little voids indicate the location of the substrate-layer-interface. No etch has been applied to delineate the interface. SEM, kindly taken by Dr. H. Lauckner, SEL Stuttgart

nism. These striations delineate the instantaneous shape of the growth interface. They may be caused by slight temperature fluctuations or by vibrations. Type A striations can be deliberately caused by applying to the growing crystals current pulses or vibrational pulses at definite time intervals ("demarcation technique"[127]). The second type of striations, named „Type B striations", is related to the growth mechanism of the layer. Type B striation in GaP layers, as well as in Se doped GaAs layers, delineate traces of the risers of surface terraces[123–125] (Fig. 34). An analysis by cathodoluminescence and spatially resolved photoluminescence[124]

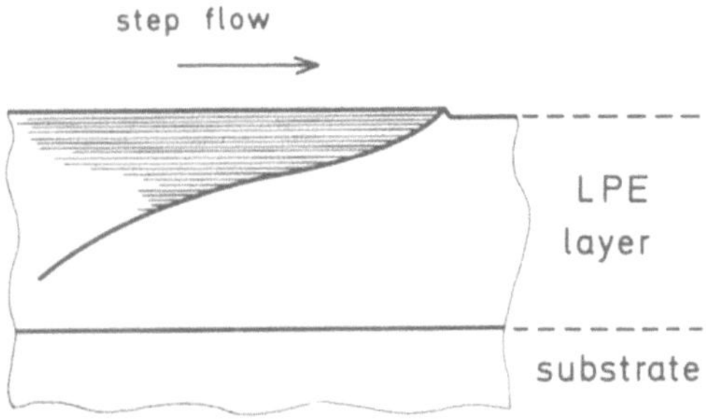

a

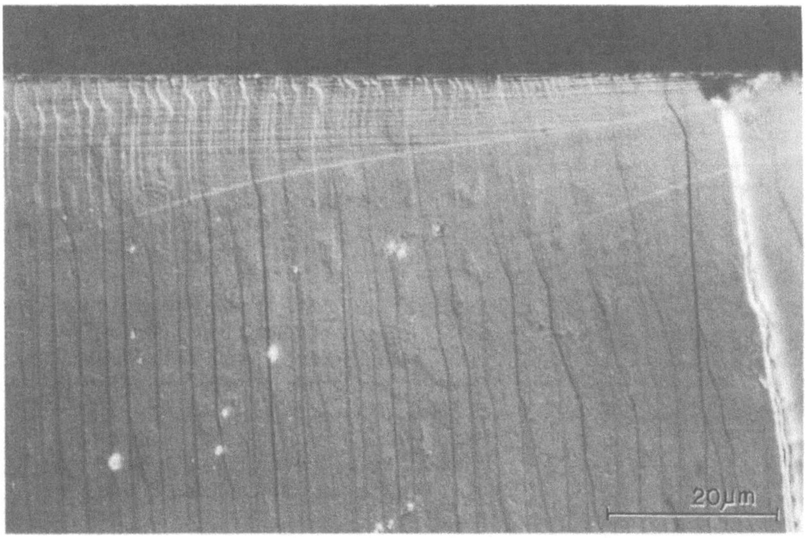

Fig. 34 a and b. Type B striations in LPE GaAs.
a Schematic of one strong line which obliquely traverses the layer. It begins near the substrate as faint line and becomes more distinct as it approaches the surface. When approaching the surface, the strong line ends there at the riser portion of a terrace. The fine lines which are approximately parallel to the terrace treads always end at the strong line
b Photoched cleavage face of a Se-doped GaAs layer showing strong and fine lines. The irregular vertical lines are unwanted cleavage steps

showed that the type B striation regions were areas of increased dopant concentration (Fig. 35). Photoetched cleavage planes revealed an additional structure of fine striations. An example of a complete type B striation pattern of a terraced, Se-doped GaAs layer is given in Fig. 34a and b. From this pattern of striations, details of the terrace growth mechanism could be deduced:
The strong lines in the etch patterns and the corresponding lines in the luminescence images obviously originate in the epitaxial layer. They begin near the substrate as faint lines and traverse the layer obliquely with various curvatures. The lines usually become more distinct and often broaden as they approach the surface of the layer.

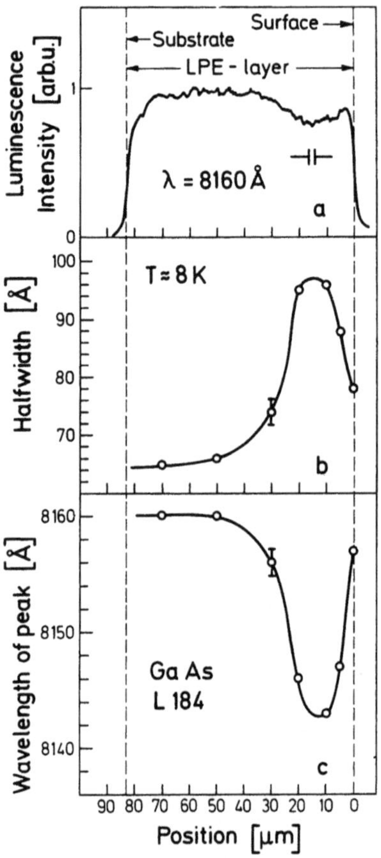

Fig. 35 a–c. Spatially resolved photoluminescence showing that the dopant Se is preferentially incorporated at the risers of terraces. after Fischer et al.[124)]

a Intensity profile across the cleavage face. A minimum appears where the exciting beam scans across a terrace trace (at 15 μm on the position scale)

b and c The half-width and the wavelength of the emission peak, as obtained by single readings, are plotted as functions of position on the cleavage face. At the positions of the minimum, the half-width is increased and the peak is shifted to smaller wavelength. Both effects indicate that the intensity minimum is caused by a region of slightly increased doping concentration (after Fischer et al.[124)])

Once such a strong line reaches the surface, it ends at the riser portion of a terrace. These strong lines have been reported to be traces of the riser portion of surface terraces[123–125)]. In addition, there is a pattern of fine lines which run approximately parallel to the treads of the surface terraces. The fine lines are not continuous, but always end at a strong line, as is shown in Fig. 34b (schematicly in Fig. 34a). These fine, parallel lines have been interpreted as traces of terraces of very small height which move on the treads of the high terraces. Incorporating the evidence from this observation, a growth model for terrace growth can be given as follows:[124, 125)]

Epitaxial growth starts from a substrate surface with misorientation steps. The heights of these steps and their mutual distances are randomly distributed. By adsorption of new molecules at their edges, the steps move laterally with a velocity approximately proportional to h^{-1}, as was reported previously[128)]. In this simplified model, steps of smaller height will catch up with higher steps, thus causing the average step height and step distance to increase. Due to this "bunching" of small growth steps, surface terraces gradually develop. As growth proceeds, these terraces become progressively immobile while increasing in height. Yet, the crystal continues to grow.

For this growth in the presence of high terraces, a lateral mechanism has been suggested by Fischer and Bauser[124, 125]. The lateral mechanism may be deduced from the pattern of fine striations in Fig. 34: small steps are generated at the base of the terrace risers and advance on the treads. The velocity of these small steps will be much greater than the velocity of the high terraces. The growth of the layer is a result of the lateral motion of both the high terraces and the small steps. As they move on the treads of the terraces, the small steps, are also subject to bunching. Thus, their height increases as they move toward the edge of the terrace, and as a result, they may cause fine striations approximately parallel to the terrace treads as indicated in Fig. 34. The contribution of high terraces to the total growth of the layer may thus be only a small percentage.

3.4 Doping Behavior and Growth Relations

Electrical transport and carrier recombination properties depend on the presence of particular dopants. In this chapter we will discuss some doping pecularities of III–V compounds connected with the layer growth.

In general, III–V compounds behave normally upon the introduction of dopants. This means that elements of the IInd Group act as acceptors when substituting the III Group Component, and that those from the VI Group act as donors when substituting the V Group component of the compound[60]. However, the group IV elements behave in a more complex fashion since they can go on either lattice site. From early theoretical considerations by Welker[1], it is obvious that in III–V semiconductors the doping element substitutes the lattice element with the greater covalent tetrahedral radius. Values for the covalent tetrahedral radii of various Group III and Group V elements are given in Table 4[129]. Si may be used as an example of this doping behavior. In those III–V compounds with similar radii of the components (compare Table 4), Si (r_{Si} = 1,173 Å), a typical amphoteric behavior,

Table 4. Doping behavior of various group-IV elements in III-V semiconductors

	Si	Ge	Sn
AlSb	A[143]	A[143]	A/D[143]
GaP	A/D	A/D[133]	D[133]
GaAs	A/D	A/D	A/D
GaSb	A[137]	A[137, 144]	A[137, 144]
InP	D[136, 142]	D[142, 144]	D[142, 144]
InAs	D[138, 145]	D[144]	D[144]
InSb	A[130, 146]	A	D[130]

A = Acceptor. D = Donor. Incorporation of Group IV elements in III-V-compounds.

appears as is observed in the case of GaAs[130, 131] and, though less pronounced, in GaP[132, 133]. If Si is incorporated into the crystal at a low growth temperature, p-type layers will result, whereas at a higher growth temperature the layers come out n-type. This ratio $[Si_{Ga}]/[Si_{As}] = 1$ at the n/p conversion temperature T_c. The value for T_c was estimated to be $T_c = 890\ °C$[134] and was shown to be dependent on the substrate orientation[135]. However, Si in InP prefers the In site due to the large difference between the atomic radii of In and P[136]. Similar conclusions can be drawn for GaSb (Si predominate acceptor[137] and InAs (Si predominate donor[138]). The other Group IV elements, Ge and Sn, have larger atomic radii than Si ($r_{Ge} = 1{,}225$ Å; $r_{Sn} = 1{,}405$ Å). Accordingly, Ge[139, 140] and Sn[139, 141] show an amphoteric behavior in GaAs. In InP, Ge and Sn should prefer to be incorporated on an In-site as a donor; this has actually been observed[142].

Influence of Dopants on Surface Morphology. Dopant species and dopant levels have a strong effect on surface terracing, exceeding that due to temperature gradients or supersaturation. Terracing on surfaces of InP LPE layers[136, 148], for instance, has been observed to be considerably reduced when certain dopants become incorporated into the crystal[147]. For InP substrates, whose orientations are 1 deg. off the (111) lattice plane, the critical doping concentration required for smooth LPE layers has been estimated[147] (Table 5). The dopants considered are Si, Ge, Zn, and Sn.

Table 5. Estimated critical doping concentrations required for smooth LPE as a function of dopant atomic radius (after Liu[147])

	Si	Ge	Zn	Sn
Atomic radius R_{dopant} (Å)	1.17	1.22	1.31	1.40
$R_{dopant} - R_{indium}$ (Å)	−0.27	−0.22	−0.13	−0.04
Critical doping level ($\times 10^{18}$ cm^{-3})	$\leqslant 4$	$\leqslant 4$	~ 4	~ 8

The models for the smoothing effect due to the presence of doping atoms are still under discussion. The findings of Kajimura et al.[123], Fischer et al.[124], and Bauser et al.[125], that for GaP and GaAs, impurities tend to segregate preferentially at the risers, may support a model described by Chernov[149]. According to Chernov, small steps are usually generated at the reentrant edges of macrosteps (terraces). The presence of dopants is likely to enhance such a generation of small steps at the base of terraces. The lateral growth mechanism observed for Se-doped GaAs by Fischer et al.[124] and Bauser et al.[125] may become more dominant. The average terrace height may thus be continuously reduced when high doping concentrations are applied. However, Liu[147] suspected that the preferential segregation of impurities with respect to terraces is a separate phenomenon unrelated to the smoothing effect caused by dopants. More work obviously needs to be done in order to fully understand the influence of impurity atoms on surface morphology.

4 Summary

We have described various methods used to grow bulk III–V semiconductors from metallic solutions. The Travelling Solvent Techniques turned out to be the most suitable growth methods for the production III–V semiconductor crystals of reasonable size and reproducible crystallographic and physical properties. The possibilities of growing thin-layer single-crystal structures by liquid phase epitaxy have been considered. Growth mechanisms in liquid phase epitaxy which influence material properties, e. g., surface morphology, dopant incorporation, and dopant homogeneity, have been described. Causes for common defects, such as terraces, striations, and voids, have been reported. The results of these studies may be applied to establish appropriate conditions in solution growth of III–V semiconductors as a further step in optimizing material properties.

5 References

1. Welker, H.: Z. Naturforschg. *7a*, 744 (1952), *8a*, 248 (1953)
2. Logan, R. M., Hurle, D. T. J.: J. Phys. Chem. Sol. *32*, 1739 (1971)
3. Van der Meulen, Y. J.: J. Phys. Chem. Sol. *28*, 25 (1967)
4. Delves, R. T.: Theory of Interface Stability in Crystal Growth (Ed. Pamplin, B. R.) Pergamon Press 1975, p. 40
5. Jakowetz, W., Rühle, W., Breuninger, K., Pilkuhn, M.: Phys. Stat. Sol. (a) *12*, 169 (1972)
6. Rado, W. G. and Crawley, R. L.: J. Appl. Phys. *43*, 1316 (1972)
7. Plaskett, T. S., Woods, F. J.: J. Crystal Growth *11*, 341 (1971)
8. Antypas, G. A.: Preparation of high-purity bulk InP. Inst. Phys. Conf. Ser. No. 33b, 55 (1977)
9. Barthruff, D., Benz, K. W., Antypas, G. A.: J. Electron. Materials *8*, 4, 485 (1979)
10. Pfann, W. G.: Trans. Met. Soc. AIME *203*, 961 (1955)
11. Broder, J. D., Wolff, G. A.: J. Electrochem. Soc. *110*, 17, 1150 (1963)
12. Wolff, G. A., Mlavsky, A. I.: Travelling Solv. Techniqu. Crystal Growth. Vol. 1 (Ed. C. H. L. Goodman) Plenum Press 1974
13. Mlavsky, A. I., Weinstein, M.: J. Appl. Phys. *34*, 9, 2885 (1963)
14. Tiller, W. A.: J. Appl. Phys. *34*, 2757 (1963)
15. Weinstein, M., Mlavsky, A. I.: J. Appl. Phys. *35*, 1892 (1964)
16. Wolff, G. A., LaBelle, H. E., Das, B. N.: Trans. Met. Soc. AIME *242*, 436 (1968)
17. Wolff, G. A., Das, B. N.: J. Electrochem. Soc. *113*, 299 (1966)
18. Joullie, A.: Thesis, Univers. Montpellier 1975
19. Weinstein, M., LaBelle, H. E., Mlavsky, A. I.: J. Appl. Phys. *37*, 2913 (1966)
20. Wilke, K. T.: Kristallzüchtung. VEB Dtsch, Verlag d. Wiss., Berlin 1973
21. Plaskett, T. S., Blum, S. E., Foster, L. M.: J. Electrochem. Soc. *114*, 1303 (1967)
22. Eyer A., Nitsche, R.: Prep. exper. growth of silicon crystals in the mirror heating facility during the first Spacelab mission. Proc. 3rd Europ. Symp. on Material Sci. in Space, Grenoble 1979 paper II.3
23. Hemmat, N. et al.: Solution growth of Electronic Compounds and their solid solutions. Proc. Mat. Eng. Sci. Div. Biennial Conf. Am. Inst. Eng. p. 112 (1970)
24. Wilcox, W. R.: National Technical Information Service U.S. Department of Commerce Report AD – 756295 (1973)
25. Benz, K. W., Müller, G.: J. Crystal Growth *46*, 35 (1979)
26. Benz, K. W. and Müller, G.: High-Quality III–V–Semiconductor Substr. Crystals Grown by THM., Proc. 7th Int. Symp. GaAs and Related Cds., Inst. Phys. Con. Ser. No. 45 p (1979)

27. Bell, R. O.: J. Electrochem. Soc. *121,* 1366 (1974)
28. Langbein, D.: Heat Flow, Diff. Convection in the Travelling Solv. Meth., Proc. 2nd Europ. Symp. Material Sci. in Space, esa SP 114, p. 223 (1976)
29. Neuberger P.: III–V–Semiconducting Cds in: Handb. of Electronic Materials, Vol. 12 IFI Plenum Press New York 1971
30. Yip, Y. F. S., Chang, C. E., Wilcox, W. R.: J. Crystal Growth *29,* 69 (1975)
31. Wald, F. V., Bell, O.: J. Crystal Growth *30,* 29 (1975)
32. Wald, F. V., Entine, G.: Nucl. Instr. Methods *150,* 13 (1978)
33. Long, S. J., Ballantyne, J. M., Eastman, L. F.: J. Crystal Growth *26,* 13 (1974)
34. Rhode, D. L.: J. Crystal Growth *20,* 13 (1973)
35. Frosh, C. I., Derich L.: J. Electrochem. Soc. *108,* 251 (1961)
36. Nguyen, Van Mau A., Ance, C., Bougnot, G.: Mater. Res. Bull. *3,* 901 (1968)
37. Yip, V. F. S., Wilcox, W. R.: J. Crystal Growth *36,* 29 (1976)
38. Benz, K. W., Müller, G.: III–V–Semiconductor Cryst. Grown by the Travelling-Heater-Method. Proc. 3rd Europ. Symp. Mater. Sci. Space, Grenoble, April 1979
39. Griffiths, L. B.: J. Phys. Chem. Sol. *27,* 257 (1966)
40. Noack, R. A., Rühle, W., Morgan, T. N.: Phys. Rev. B *18,* 6944 (1978)
41. Kaneko, K. et al.: Proc. Institute of Electrical and Electronics Engineers *61,* 884 (1973)
42. Roksnoer, P. J. et al.: J. Crystal Growth *40,* 6 (1977)
43. Gillessen, K., Marshall, A. J.: J. Crystal Growth *32,* 216 (1976)
44. Moravec, F., Novotny, J.: J. Crystal Growth *33,* 90 (1976)
45. Gillessen, K., Marshall, A. J.: J. Crystal Growth *33,* 356 (1976)
46. Rodot, H., Hruby, A., Schneider, M.: J. Crystal Growth *3,* 305 (1968)
47. Rosgonyi, G. A., Afromowitz, M. A.: App. Phys. Lett. *19,* 153 (1971)
48. Kaneko, K., Ayabe, M., Isawa, N.: Jap. J. Appl. Phys. *18,* 861 (1979)
49. Marshall, A. J., Gillessen, K.: J. Crystal Growth *44,* 651 (1978)
50. Ito, K., Ito, H.: J. Crystal Growth *45,* 248 (1978)
51. Hsieh, J. J.: Thickness of InP layers grown by LPE from supercooled solutions. GaAs and Related Cds. 1976, Inst. Phys. Conf. Ser. 336, p. 74
52. Gillessen, K., Marshall, A. J., Hesse, J.: Temperature Gradient Solution Growth; Application 1979, p. to III–V Semiconductors. Crystals *3,* Berlin: Springer
53. Nelson, H.: RCA Rev. *24,* 503 (1963)
54. Dawson, L. R.: Liquid phase epitaxy. In: Prog. in Sol. State Chem. 7 (Reiss, H., McCaldin, H. O., Eds.) Oxford: Pergamon 1972, p. 117
55. Kressel, H., Nelson, H.: Properties and applications of III–V cds. films deposited by liquid phase epitaxy. In: Physics of Thin Films, Vol. 7. New York: Academic Press 1973, pp. 115
56. Giess, E. A. Ghez, R.: Liquid phase epitaxy. In: Epitaxial Growth. Part A (Matthews, J. W. Ed.) New York: Academic Press 1975, p. 183
57. Deitch, R. H.: Molten metal solution growth. In: Crystal Growth *6.* Oxford: Pergamon 1975, p. 427
58. Brice, J. C.: In: Curr. Top. Mater. Sci. *2* (Kaldis, E. Ed.) Amsterdam: North Holland 1977, pp. 571
59. Randles, M. H.: Liquid phase epitaxial growth of magnetic garnets. In: Crystals 1. Berlin: Springer 1978, p. 71
60. Zschauer, K. H.: Liquid Phase Epitaxy. Proc. Europ. Summer School, Bad Boll 1974, p. 115
61. Foster, L. M.: The Preparation of III-V Compounds Semiconductor Alloys. In: Preparation and Properties of Solid State Material, Vol. 3. (Eds. Wilcox, W. R., Lefever, R. A.) New York and Basel: Marcel Dekker 1977
62. Rupprecht, H., Woodall, J. M., Pettit, G. D.: Appl. Phys. Lett. *11,* 81 (1967)
63. Grobe, E., Salow, H.: Z. Angew. Physik *32,* 381 (1972)
64. Lorimor, O. G. et al.: Sol. State Electronics, *16,* 1289 (1973)
65. Saul, R. H., Lorimor, O. G.: J. Crystal Growth *27,* 183 (1974)
66. Mottram, A., Peaker, A. R.: J. Crystal Growth *27,* 193 (1974)
67. Alferov, Zh. I. et al.: Soviet Phys. Semiconductors *2,* 1289 (1969)
68. Panish, M. B., Hayashi, I., Sumski, S.: Appl. Phys. Lett. *16,* 326 (1970)

69. Woodall, J. M.: J. Crystal Growth *12*, 32 (1972)
70. Thompson, G. H. B., Kirkby, P. A.: J. Crystal Growth *27*, 70 (1974)
71. Hayashi, I. et al.: Appl. Phys. Lett. *17*, 109 (1970)
72. Woodall, J. M., Hovel, H. J.: Appl. Phys. Lett. *21*, 379 (1972)
73. Small, M. B., Blackwell, J. C., Potemski, R. M.: J. Crystal Growth *46*, 253 (1979)
74. Scheel, H. J.: J. Crystal Growth *42*, 301 (1977)
75. Dawson, L. R., Whelan, J. M.: Bull. Amer. Phys. Soc., Series 2, *13*, 375 (1968)
76. Donahue, J. A., Minden, H. T.: J. Crystal Growth *7*, 221 (1970)
77. Lockwood, H. F., Ettenberg, M.: J. Crystal Growth *15*, 81 (1972)
78. Panish, M. B., Sumski, S., Hayashi, I.: Met. Trans. AIME *2*, 795 (1971)
79. Akita, K. et al.: J. Crystal Growth *35*, 279 (1976)
80. Bauser, E. et al.: Japan J. Appl. Phys. *16*, Suppl. 16–1, 457 (1977)
81. Kressel, H.: J. Electron. Mater. *3*, 747 (1974)
82. Lang, D. V. et al.: J. Appl. Phys. *47*, 2558 (1976)
83. Kossel, W.: Die Naturwissenschaften, *18*, 131 (1930)
84. Strickland-Constable, R. F.: Kinetics and Mechanism of Crystallisation. London: Academic Press 1968
85. Pechlyakov, O. P. et al.: Phys. Stat. Sol. (a) *17* (1973) 339
86. Cabrera, N., Coleman, R. V.: Theory of Crystal Growth from the Vapor. In: The Art and Science of Growing Crystals (Gilman, J. J. Ed.) New York: Wiley 1966, pp. 3
87. Bennema, P. Gilmer, G. H.: Kinetics of crystal growth. In: Crystal growth: an introduction (Hartman, P. Ed.) Amsterdam: North Holland 1973, pp. 263
88. Peters, R. C.: Growth Characteristics and Morphology of GaP Liquid Phase Epitaxy. Proc. Fourth Intern. Symp. on Gallium Arsenide, Inst. of Physics, Conf. Ser. No. 17, 1973, p. 55
89. Bauser, E. et al.: J. Crystal Growth *27*, 148 (1974)
90. Saul, R. H., Roccasecca: J. Appl. Phys. *44*, 1983 (1973)
91. Rode, D. L.: Phys. Status Solidi (a) *32*, 425 (1975)
92. Mattes, B. W., Route, R. K.: J. Crystal Growth *27*, 133 (1974)
93. Tiller, W. A.: J. Crystal Growth *2*, 69 (1968)
94. Donahue, J. A., Minden, H. T.: J. Crystal Growth *7*, 221 (1970)
95. Crossley, I., Small, M. B.: J. Crystal Growth *19*, 160 (1973)
96. Vilms, J., Garrett, J. P.: Sol. State Electronics *15*, 443 (1972)
97. Longo, J. T. et al.: J. Crystal Growth *15*, 107 (1972)
98. Minden, H. T.: J. Crystal Growth *6*, 228 (1970)
99. Small, M. B., Barnes, J. F.: J. Crystal Growth *5*, 9 (1969)
100. Hsieh, J. J.: J. Crystal Growth *27*, 49 (1974)
101. Toyoda, N., Mihara, M., Hara, T.: Appl. Phys. Lett. *27*, 627 (1975)
102. Deitch, R. H.: J. Crystal. Growth *7*, 69 (1970)
103. Moon, R. L., Vander Plas, H. A.: J. of Electron. Mater. *7*, 347 (1978)
104. Messham, R., Majerfeld, A.: J. Electronic Mater. *9*, to be published
105. Rode, D. L.: J. Crystal Growth *27*, 313 (1974)
106. Bauser, E. et al.: Verh. Dtsch. Physik. Ges. *7*, 505 (1976)
107. Bruni, H. J.: Gadolinium Gallium Garnet. In: Crystals *1*. Berlin: Springer 1978, p. 53
108. Bauser, E.: Steps on growth facets of GaAs epitaxial layers. To be published
109. Frank, F. C.: Faraday Soc. *5*, 48 (1949)
110. Burton, W. K., Cabrera, N., Frank, F. C.: Phil. Trans. Roy. Soc., London *243*, 299 (1951)
111. Bauser, E., Hagen, W.: X-ray topography of growth-step sources in LPE Gallium Arsenide. Submitted to J. Crystal Growth
112. Bauser, E. Strunk, H.: to be published
113. Linnebach, R. et al.: to be published
114. Kuphal, E.: to be published
115. Small, M. B., Bachem, K. H., Potemski, R. M.: J. Crystal Growth *39*, 216 (1977)
116. Small, M. B. et al.: J. Crystal Growth *30*, 257 (1975)
117. Stringfellow, G. B.: Private communication
118. Bauser, E.: Appl. Phys. *15*, 243 (1978)

119. Lorenz, M. R., Pilkuhn, M.: J. Appl. Phys. *37,* 4094 (1966)
120. DeKock, A. J. R. et al.: J. Crystal Growth *4,* 13 (1977)
121. Carruthers, J. R., Witt, A. F.: Transient Segregation effects in Czochralski growth. Lecture Notes, Sec. Internat. Spring School Crystal Growth, 1974 Japan, L9
122. Kim, K. H.: J. Elchem. Soc. *126,* 875 (1979)
123. Kajimura, T., Aiki, K., Umeda, J.: Appl. Phys. Lett. *30,* 526 (1977)
124. Fischer, B. et al.: Appl. Phys. Lett. *33,* 78 (1978)
125. Bauser, E., Fischer, B., Sullivan, P. A.: Tracing of terrace growth in GaAs LPE layers. In: Semiconductor Characterization Techniques. (Barnes, P. A., Rozgonyi, G. A. Eds.) Electronic Div. Electrochemical Soc., Proc. *78–3,* Princeton 1978
126. Faust jr., J. W., John, H. F., O'Hara, S. In: Metallurgy of Seminconductor Materials (Schroeder, J. B. Ed.) New York, Interscience: 1962, p. 181
127. Singh, R., Witt, A. F., Gatos, H. C.: J. Electrochem. Soc. *115,* 113 (1968) Received October
128. Chernov, A. A.: Usp. Fiz. Nauk *73,* 277 (1961) Sov. Physics Uspekhi *4,* 116 (1961)
129. Hübner, K.: Kristall u. Technik *9,* 865 (1974)
130. Edmond, J. T.: Proc. Phys. Soc. *73,* 622 (1959)
131. Ashen, D. J. et al.: Phys. Chem. Sol. *36,* 1041 (1975)
132. Lorenz, M. R., Pilkuhnn, M. H.: J. Appl. Phys. *38,* 61 (1967)
133. Dean, P. J., Frosh, C. J., Henry, C. H.: J. Appl. Phys. *39,* 5631 (1968)
134. Teramoto, I.: J. Phys. Chem. Sol. *33,* 2089 (1972)
135. Ahn, B. H., Schurtz, R. R., Trusell, C. W.: J. Appl. Phys. *42,* 4512 (1972)
136. Baumann, G. G., Benz, K. W., Pilkuhn, M.: J. Electrochem. Soc. *123,* 1232 (1976)
137. Jakowetz, W., Barthruff, D., Benz, K. W.: High-purity and doped GaSb layers grown by LPE and VPE techniques. GaAs and Related Cds. 1976, Inst. Phys. Conf. Ser. No. 33a, p. 41
138. Mizuno, O., Watanabe, H., Shinoda, D.: Japan. J. Appl. Phys. *14,* 184 (1975)
139. Summers, C. J., Dingle, R., Hill, D. E.: Phys. Rev. B *1,* 1603 (1970)
140. Ploog, K.: Molecular Beam Epitaxie of III-V Compounds Crystals *3,* 73 (1980)
141. Schairer, W., Grobe, E.: Sol. State Comm. *8,* 2017 (1970)
142. Astles, M. G., Smith, F. G. H., Williams, E. W.: J. Electrochem. Soc. *120,* 1750 (1973)
143. Shaw, D.: Proc. Phys. Soc. *80,* 161 (1962)
144. Folbert, O. G., Schillmann, E.: Z. Naturforschung *12a,* 943 (1957)
145. Schillmann, E.: Z. Naturforschung *11a,* 463 (1956)
146. Cunningham, R. W.: Int. Conf. Phys. of Semiconductors Proc. Exeter, 732 (1962)
147. Liu, Y. Z.: The effect of doping species and doping levels on the terracing of LPE growth of InP on misoriented substrates, Gallium Arsenide and related compounds (Wolfe, C. M. Ed.) Inst. Phys. Conf. Ser. No. 45, 103 (1978)
148. Pak, K., Nishinaga, T., Uchiyama, S.: Japan. J. Appl. Phys. *16,* 949 (1977)
149. Chernov, A. A.: Surface morphology and growth kinetics. ISSCG2, Sec. Internat. Spring School Crystal Growth, Japan 1974

Received October 18, 1979

Temperature Gradient Solution Growth
Application to III-V Semiconductors

Klaus Gillessen, Albert J. Marshall, and Joachim Hesse

AEG-Telefunken, Semiconductor Division, D-7100 Heilbronn, Federal Republic of Germany

This article reviews the Temperature Gradient Solution (TGS) growth method (sometimes also named Synthesis, Solute Diffusion technique) of preparing bulk crystals of III-V semiconductors. This method is an interesting alternative to the conventional Liquid Encapsulation Czochralski (LEC) technique because the properties of this material are similar to those of epitaxial layers. Therefore, production of devices is possible without using liquid or vapour phase epitaxy. The principle of the method is described using GaP as an example. Experimental techniques for growing single crystals with controlled doping are thoroughly discussed. Growth of other binary compounds (e.g. InP) is also described. A variation of the TGS method can be applied to the growth of ternary compounds (mainly $In_{1-x}Ga_xP$). The defect densities and electrical properties of TGS grown crystals are compared to the corresponding properties of LEC material. Preparation and properties of LED's based on TGS grown materials are treated in detail. With further development work the TGS growth method may become competitive with the standard production process (LEC plus epitaxy) for LED materials.

Table of Contents

1 Introduction

Crystals of III-V semiconductors have found widespread applications. They are the basis for devices such as light emitting diodes (LED's), semiconductor lasers, photodetectors, high frequency transistors and diodes, Gunn effect devices, Hall generators, and photocathodes. To realize all of these devices high quality material is required. This means that large single crystals with a low defect density and well controlled impurity concentrations must be available. Due to their tendency to decompose at elevated temperatures, the III-V compounds are generally more difficult to grow in single crystalline form than the elemental semiconductors germanium and silicon. The two methods most frequently applied are the horizontal Bridgman growth for compounds with low or medium vapour pressure (e.g., GaAs) and the liquid encapsulated Czochralski (LEC) technique for compounds with high vapour pressure (e.g., GaP and InP). Both methods are now highly perfected and yield crystals of considerable size: diameters of ~50 mm are standard, ~76 mm slices are also available. However, the structural perfection of these crystals which strongly influences important semiconductor parameters like minority carrier lifetime and mobility, is still insufficient for most device applications. Therefore, the melt grown crystals are only used as substrates, and epitaxial techniques are generally employed with III-V compounds to produce material with higher quality for the active region of a device. The most important difference between bulk crystal growth and vapour or liquid phase epitaxy is that the latter takes place at considerably lower temperatures than the former. This decreases the number of point defects introduced during growth, alleviates the control of undesirable impurities, and diminishes the elastic stresses introduced during cooling to room temperature. A crystal growth method operating at low epitaxy temperatures and yielding large sized crystals as melt growth techniques would be able to replace both steps and thus simplify III-V technology very much. This review describes a possibility to achieve this aim: the Temperature Gradient Solution Growth (TGS). (Another term for this crystal growth method is Synthesis, Solute Diffusion (SSD) technique. However, this nomenclature is not generally accepted.) Because this method was first applied to gallium phosphide and is up to now in most cases used for crystal growth of this compound, this review describes the basic ideas and features using GaP as an example. Applications to other III-V compounds (InP, $In_{1-x}Ga_xP$) are also summarized.

2 Crystal Growth

2.1 Principle and Comparison with Related Methods

If the solubility of a substance in a solvent increases with temperature as it is the case for a large class of combinations, the substance can be grown from solution by two basic methods:

1) The solution is saturated at a particular temperature and cooled to a lower temperature (temperature programme method). As the solubility decreases with falling temperature the substance is forced to precipitate.

2) The solution is held in a temperature gradient with the hotter end saturated with the substance (temperature gradient method). Here the solute is transported by diffusion or convection plus diffusion to the cooler end and solidified due to the smaller solubility.

The temperature programme method is widely used for liquid phase epitaxy (LPE). It is less suitable for the growth of bulk crystals simply because the amount of material which can be grown in one process is very limited. A second less obvious disadvantage is that the growth temperature is varied through a large interval which will result in very inhomogeneous crystal properties (e.g., vacancy concentrations, impurity incorporation). With the temperature gradient method the growth temperature can be kept constant, and in principle large crystals can be grown with a limited amount of solvent. In the case of gallium phosphide (and with most other III-V compounds, too) the most obvious choice of a solvent is the group III component, gallium, because this avoids the problem of impurity incorporation from a "hetero" solvent. Furthermore, it has a low melting point, low vapour pressure, and is available in high purity. The binary phase diagram gallium-phosphorus[1)] shown in Fig. 1 has to be considered to understand the principle of the TGS method. The liquidus on the gallium-rich side gives the solubility of phosphorus in gallium at a given temperature. For example, at T = 1120 °C the phosphorus solubility is about 5 at %. The solidus is practically a vertical line at 50% phosphorus, i. e., the solid precipitated from saturated gallium-phosphorus solutions is the stoichiometric composition GaP, independent of temperature (or liquidus composition). At temperatures close to the melting point measurable deviations from stoichiometry exist[2)], however, they can be neglected. Another important point is the pressure and vapour composition in equilibrium with the liquidus composition[1)], which is shown in Fig. 2: Along the gallium-rich side of the liquidus the vapour consists mainly of P_2,

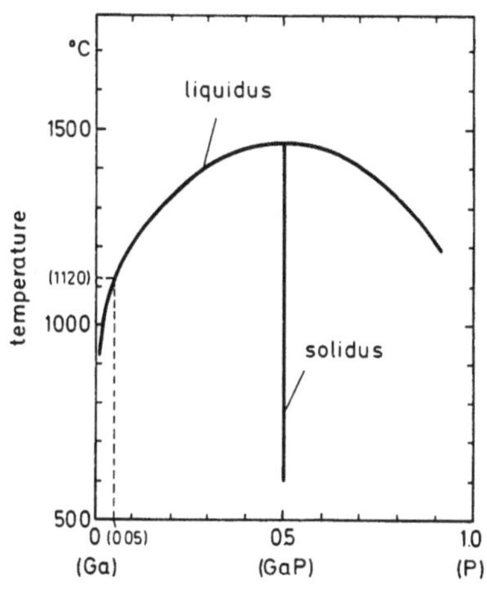

Fig. 1. Gallium-phosphorus phase diagram

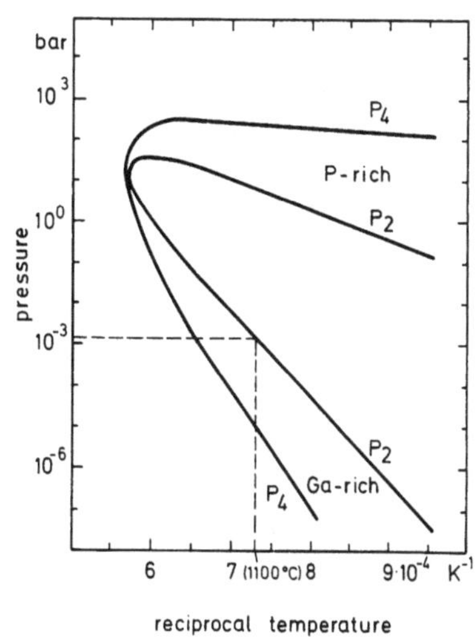

Fig. 2. Vapour pressures along the liquidus in the Ga-P system

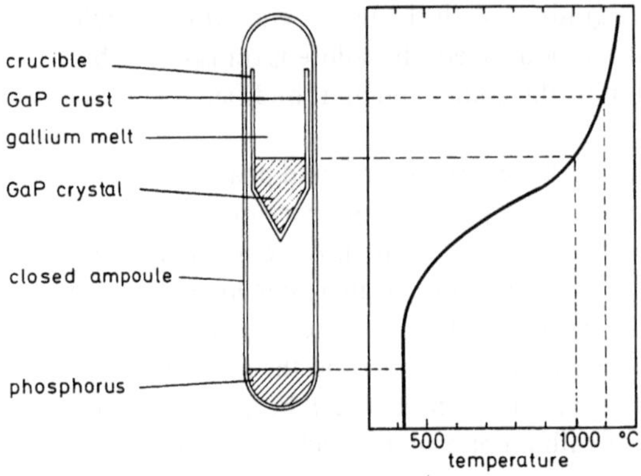

Fig. 3. Principle of the temperature gradient solution (TGS) growth method

and the equilibrium pressure at 1100 °C is around 10^{-3} bar. This means that if a gallium melt at 1100 °C is brought into contact with phosphorus vapour with a pressure larger than 10^{-3} bar, a gallium phosphide crust will develop. With this basic knowledge of the gallium-phosphorus phase diagram one can now understand the principle of the TGS growth method, which is schematically shown in Fig. 3: Initially, a gallium melt contained in a crucible is placed in a temperature gradient so that the bottom of the melt is typically at around 1000 °C and the surface is at about 1100 °C. The crucible is fixed inside a closed ampoule which contains elemental phosphorus at its lower end. This phosphorus reservoir is usually heated to 420 °C where the phosphorus vapour pressure is just one atmosphere. Because the lower end is the coldest part of the system, the whole ampoule volume is filled with phosphorus vapour of atmospheric pressure. At the exposed gallium surface a gallium phosphide crust is formed which is continuously being dissolved by the gallium melt. Phosphorus is transported by diffusion (or convection plus diffusion) to the bottom of the crucible, where the temperature is lower, the solubility of phosphorus in gallium is consequently smaller, and gallium phosphide is precipitated. This process – phosphorus evaporation, crust formation and dissolution, solute transport, gallium phosphide precipitation – continues until either the phosphorus reservoir or the gallium melt is exhausted. This crystal growth method was first described in 1968 and applied to GaP by Rodot, Hruby, and Schneider[3)] and compared with the Bridgman method by Besselere and Le Duc[4)]. A thorough treatment of the technique is given by Kaneko et al.[5)] who also introduced the denomination Synthesis, Solute Diffusion (SSD). At almost the same time Poiblaud and Jacob[6)] published similar results. Gillessen and Marshall[7)] succeeded in 1976 to reproducibly grow single crystals by the SSD method. Moravec and Novotny[8)] studied the influence of several growth parameters on the grain size of SSD-GaP.

To differentiate the TGS or SSD growth technique from related methods its features are once more summarized:

1) a nonstoichiometric melt,
2) a temperature gradient,
3) pressure of a volatile constituent.

For example the Travelling Heater Method as applied to GaP by Broder and Wolff[9] and Plaskett, Blum and Foster[10] also uses a nonstoichiometric melt and a temperature gradient. However, the phosphorus is supplied from a solid GaP source and the temperature gradient is applied in a different manner, so that the process continues only as long as the heater is moved. The Vertical Gradient Freeze (VGF) method under phosphorus pressure was described by Blum and Chicotka[11] and Blum, Chicotka and Bischoff[12]. The difference compared to the TGS method is that here the melt is stoichiometric. The phosphorus pressure is only necessary to prevent dissociation of the melt. The liquid encapsulation vertical gradient freeze (LE-VGF) technique was applied to GaP by Blum and Chicotka[13] and Woodbury[14]. In this case the temperature gradient is the only common feature. Further methods which show some similarities to the TGS Growth are the LEC method with nonstoichiometric melts (von Neida, Oster and Nielsen[15]) and the zone melting of GaP (Frosch and Derick[16]) with a stoichiometric melt, which is otherwise similar to the Travelling Heater Method.

2.2 Gallium Phosphide

2.2.1 Materials and Apparatus

The starting materials necessary for growth of III-V compounds by the TGS method are commercially available. In the case of GaP gallium metal and elemental phosphorus (red modification) are supplied by several companies in 6N quality. Quartz is used as the crucible material by most authors, because it can be shaped easily. However, it is wetted by liquid gallium and contaminates the melt with silicon (see Sect. 3.2). Crucibles made from glassy carbon as used in Ref.[7] offer the advantage that they are not wetted by the melt. Consequently, the crystal can be separated easily from the crucible thereby allowing it to be reused several times. Some carbon contamination is introduced by this material (see Sect. 3.2). From the purity point of view pyrolytic boron nitride is an excellent crucible material[17]. However, it is quite expensive and available only in simple, thin walled shapes. Typical crucible sizes are 10 to 20 mm internal diameter. In Ref.[5] the use of crucibles up to 47 mm diameter is reported.

The phosphorus vapour is usually contained in a closed ampoule made from quartz, previously cleaned, evacuated, outgassed, and sealed by standard methods. Before sealing the ampoule the gallium melt can be outgassed at 600 to 800 °C to remove gallium oxide via the reaction:

$$Ga_2O_3 + 4\,Ga \longrightarrow 3\,Ga_2O \text{ (volatile)}.$$

The red phosphorus is normally loaded into the ampoule without further purification. In Ref.[5] the use of an open ampoule with a kind of backflow condenser is mentioned. However, this modification will certainly give rise to severe safety problems. This is most probably the reason why the closed ampoule version is generally preferred.

The temperature distribution which is used for effecting the temperature gradient solution growth consists of two regions: a low temperature zone at 420 °C with flat profile for producing the phosphorus vapour and a high temperature zone with a temperature gradient for heating the crucible, as shown in Fig. 3. It is usually produced by a vertical two-zone resistance furnace with separate thermocouple operated temperature controllers. In some cases[3, 4, 18] the low-temperature zone is placed on top of the high-temperature zone. Other authors[5, 7, 8] prefer the reversed arrangement. With this latter version it is easier to maintain a stable temperature distribution, because there is less convection within the ampoule. Poiblaud and Jacob[6] apparently used both arrangements without mentioning any advantages/disadvantages of one version against the other. In Ref.[7] induction heating is applied to produce the high temperature zone. With this method a higher temperature gradient can be achieved than with resistance heating, resulting in a higher growth speed (see Chap. 2.2.2). The temperatures of both zones should be well stabilized, because the phosphorus vapour pressure and the solubility of phosphorus in gallium are both exponentially temperature dependent. Around 420 °C the phosphorus vapour pressure varies about 2%/°C and the solubility of phosphorus in gallium changes at 1100 °C approximately 1%/°C. Some kind of overtemperature protection should be provided also, because a temperature rise of 50 °C (e.g., due to controller failure) at the phosphorus reservoir will produce an overpressure of several atmospheres, which will probably destroy the ampoule in the process. Hot phosphorus vapour ignites spontaneously when it comes in contact with air. When the growth process is finished and the ampoule cooled to room temperature, the remaining phosphorus vapour condenses within the ampoule in the form of droplets of the white modification. When the ampoule is opened this white phosphorus is exposed to air and immediately starts fuming or burning. It is therefore advisable to break the ampoule in the open air. Excess phosphorus is best ignited to produce phosphorus pentoxide P_2O_5. This compound is highly hygroscopic, and its solution in water (H_3PO_4) is nontoxic.

If a gallium phosphide crystal grows at the bottom of a crucible containing a gallium-phosphorus solution, the solid-liquid interface is shifted upwards. Because the crucible is placed in a temperature gradient (see Fig. 3) this means that the interface moves in the direction of increasing temperature. This effect is undesirable because the growth temperature is changed and the melt comes gradually into a region with a lower temperature gradient, thus decreasing the growth speed. To compensate for this effect the ampoule can be lowered with the growth speed. For this purpose a gear motor with a threaded spindle[7] or with a thin wire being unwound from a roller[3] can be used. In Ref.[4] the motion is accomplished hydraulically whereas in Ref.[6] a float resting on a water basin moves downwards as the water is evaporated slowly. For these mechanisms to work correctly the growth speed must be known before the experiment is commenced. As this is sometimes difficult, two possibilities were proposed[19] to move the crucible within the ampoule with the actual growth speed. This means that the growth-front temperature is kept constant automatically. The first method takes advantage of the fact that the weight of the crucible content increases during growth because of the incorporation of phosphorus. Its principle is shown in Fig. 4. The crucible assembly rests on a float which is immersed in a bath

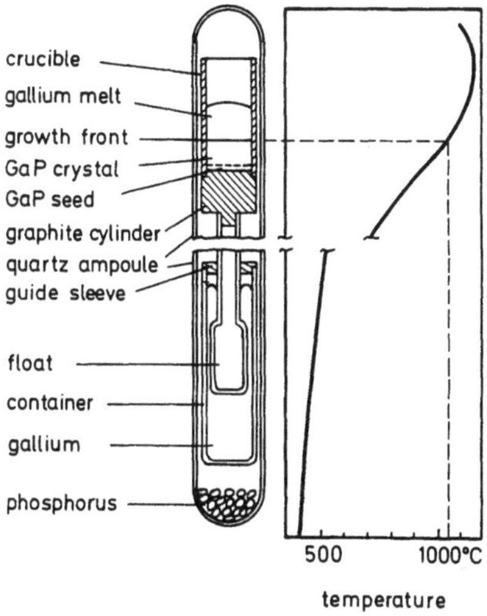

Fig. 4. Weight gain method[19)]

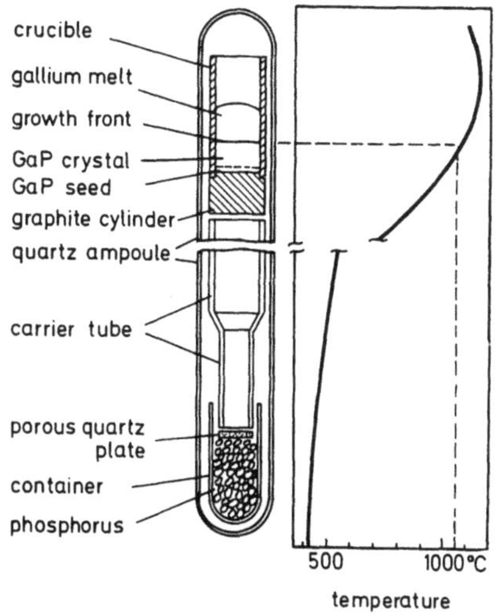

Fig. 5. Phosphorus consumption method[19)]

of liquid gallium (in principle any inert liquid could be used for that purpose). The dimensions of the float can be chosen so that the crucible is lowered at a rate corresponding exactly to the growth rate of the crystal. The second possibility utilizes the phosphorus consumption during crystal growth and is shown in Fig. 5. Here the crucible rests on the phosphorus reservoir by means of a carrier tube and a porous quartz plate. The cross section of the phosphorus container can be chosen so that the crucible motion is synchronous with the actual growth rate. Both methods for automatic maintainance of the growth front temperature have been successfully tested. However, with some experience concerning growth speeds the simpler technique of moving the whole ampoule is to be preferred because of its greater reliability. Summarizing then, the apparatus necessary for TGS growth is quite simple. This method can, therefore, easily be performed on a laboratory as well as an industrial scale.

2.2.2 Growth Process

Generally speaking, the growth speed which can be achieved with the TGS growth technique is quite low: typical values are around 1 mm per day. The reason for this undesirable feature is that the phosphorus has to be transported through the gallium melt mainly by a diffusion process. In principle, there are three possibilities to increase the flux of phosphorus atoms which is proportional to the growth speed:

1) Increase of the phosphorus concentration by choosing a higher growth temperature,
2) Increase of the concentration gradient by applying a higher temperature gradient,
3) Increase of the transport rate by introducing convection within the melt, so that diffusion takes place only through a thin boundary layer.

All three methods have been applied in practice. Kaneko et al.[5] achieved growth speeds of nearly 7 mm/day by heating the melt to temperatures in excess of 1300 °C. However, at these temperatures closed quartz ampoules can no longer be used because quartz starts to soften above 1100 °C. The disadvantage of open ampoules has been mentioned in Chap. 2.2.1. The second possibility was used in Ref.[7] where induction heating was applied to create the high temperature zone. With this method a temperature gradient within the melt of 120 °C/cm was realized, whereas in resistance heated systems values of around 40 °C/cm are typical. The increased temperature gradient resulted in high growth speeds up to 4 mm/day at moderate growth temperatures (1000–1050 °C). The growth temperature and the temperature gradient cannot be increased simultaneously if a certain maximum temperature may not been exceeded, e.g., to avoid softening of the quartz ampoule. Therefore, the position of the ampoule within a temperature profile must be carefully optimized to achieve the highest possible growth rate. This problem was treated in detail in Ref.[19], where a simple method based on temperature measurements within a gallium melt or within an empty ampoule is described to determine the position of maximum growth rate. The third possibility to increase the growth speed is to introduce convection within the melt. This can only be achieved with some kind of technical trick because the melt is contained in a closed ampoule. One possible method is the use of a stirrer with magnetic coupling to its driving mechanism. A more elegant method is to apply the accelerated crucible rotation technique (ACRT) described by Scheel[20]. When this technique is transferred to the TGS growth[17] it is easiest to rotate the whole ampoule with varying speed. The application of ACRT resulted in a moderate increase in growth rate: 2 mm/day instead of 1.5 mm/day with otherwise identical conditions. Therefore, some convection apparently takes place within the melt even without intentional stirring. The diffusion process of phosphorus in the gallium melt during the temperature gradient solution growth of GaP was studied in detail by Kaneko et al.[21]. A similar model calculation was also presented by Moravec and Novotny[8]. The main results of Kaneko's study are:

1) The diffusion coefficient of phosphorus in gallium is around 10^{-4} cm^2/s in the temperature range 1100–1200 °C, in agreement with independent measurements by other authors.

2) Between 1050 °C and ~ 1300 °C the diffusion coefficient D is only weakly temperature dependent and given by $D = 1.8 \times 10^{-2} \exp(-0.65 \text{ eV}/kT)$ cm^2 s^{-1}.

3) With a melt depth of 3 cm the time to reach a steady state is of the order of 10^5 s (approximately 1 day).

The significance of the last result is that it takes about 1 day after the start of a run until the actual growth begins.

The low growth rate of the TGS growth technique is a distinct disadvantage which apparently cannot be easily improved. However, it is partly compensated by the low cost of the equipment necessary for this method. Therefore it is possible to operate several systems in parallel, if the amount of material produced by one system is not sufficient.

2.2.3 Growth of Single Crystals

For most device applications single crystalline material is required. Therefore, some means of nucleation control has to be provided with the TGS growth technique. Most authors use pointed crucibles to initiate single crystalline growth[3, 4–6, 8, 18]. However, single crystals are only occasionally obtained with this method[3, 6]. Seeding attempts were reported by Kaneko et al.[5] and Menninger et al.[18]. In Ref.[5] the use of a thick single-crystal wafer resulted in a partial success, whereas in Ref.[18] even in seeded growth attempts (without further details being given) only polycrystalline ingots were obtained. Single crystals could be grown reproducibly with the seed configuration shown in Fig. 6[7] which was already included in Fig. 4 and 5. It consists of a disc-shaped seed crystal with its (111 B) side facing the melt and a graphite cylinder below the seed which improves the control of the heat flow. It is necessary that there is good thermal contact between the seed and the graphite cylinder, otherwise the heat flow is interrupted. If this is not the case, the temperature of the seed will increase, resulting in its dissolution. The same undesirable effect happens if the melt is not presaturated by addition of the appropriate amount of GaP pieces. The crucial point in growth of single crystals by the temperature gradient solution technique is the exact control of the starting conditions, i.e., prevention of either seed dissolution or spontaneous nucleation on its surface. Typical grain sizes of GaP polycrystals grown by this method are a few mm to 1 cm. Single crystals of 19 mm diametre were reported in Ref.[7]. An example for a single crystal is shown in Fig. 7. This value was improved to 31 mm later[17]. Figure 8 is a photograph of an SSD single crystal of this size. Even these single crystals show small grains with random orientation at their surface, apparently caused by nuclea-

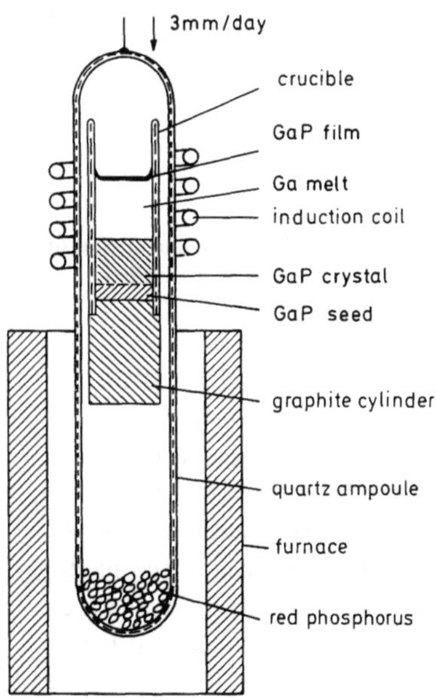

Fig. 6. TGS growth of GaP single crystals[7]

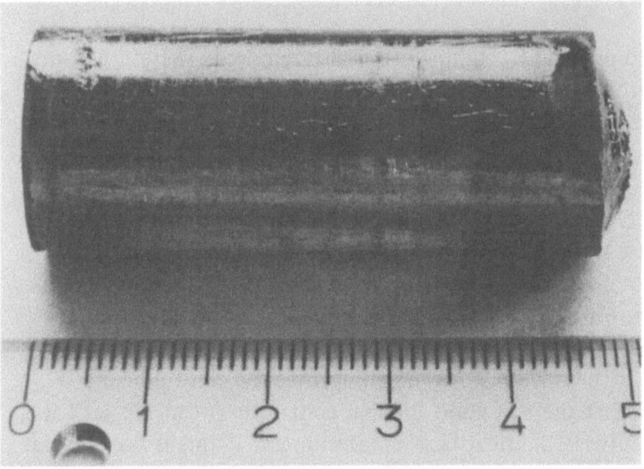

Fig. 7. TGS-GaP single crystal

Fig. 8. Single crystal of 31 mm diametre

tion at the crucible walls. This effect can be minimized by proper control of the shape of the solid-liquid interface which should be slightly convex. Note the convex shape of the crystal ends in Figs. 7 and 8. Formation of solvent inclusions is a frequent problem with solution growth techniques. This can be explained by the concept of constitutional supercooling[4, 6]. In practice this means that only below a certain growth speed or above a certain temperature gradient inclusion free material can be obtained. This behaviour was found experimentally[7], though others[8] have observed the opposite effect (dendritic growth and formation of inclusions at high temperature gradients).

2.2.4 Doping

GaP crystals can be doped n-type during the temperature gradient solution growth by addition of appropriate amounts of either elemental S, Se, Te or the corresponding gallium compounds Ga_2S_3, Ga_2Se_3, Ga_2Te_3 to the melt. According to Ref.[5] the distribution coefficient for sulphur is about 1, whereas in[17] a value of 4 was found. A possible explanation for this discrepancy is that in the latter case carbon crucibles were used, resulting in sulphur loss by formation of CS_2. In the case of tellurium the distribution coefficient varies between 0.04[5, 8] and 0.1[17], depending on the growth temperature.

Doping with zinc and oxygen (usually in the form of Ga_2O_3) which is necessary for red-light emitting material, appears to be difficult[5] because both dopants are depleted from the melt at the growth temperatures employed. Consequently elemental zinc and Ga_2O will condense at colder parts of the ampoule.

For green light emitting diodes nitrogen doped material is required. The optimum doping is close to the solubility limit which is approximately 2×10^{18} cm^{-3}. With GaN addition to the melt[7, 8, 18] some nitrogen doping can be achieved. However, the distribution coefficient of nitrogen is about 13[22], so that the nitrogen source is rapidly exhausted. In Fig. 9 the results of numerous nitrogen doping experiments with different nitrogen sources are summarized[17, 22]. With the addition of gaseous ammonia to the growth ampoule the same nitrogen distribution as in the case of GaN was found (curve 1). The N concentration falls below 10^{16} cm^{-3} within the first centimetre of crystal length and approaches a value of 4×10^{15} cm^{-3}. Nearly the same concentration (3×10^{15} cm^{-3}) was found in undoped crystals and in crystals grown in BN crucibles (curve 2). With molecular nitrogen N_2 a homogeneous doping concentration of 4×10^{16} cm^{-3} was achieved (curve 3). The same result was obtained using mixtures of nitrogen and hydrogen with an osmium catalyst to accelerate the formation of ammonia. Hardly any ammonia was formed with this

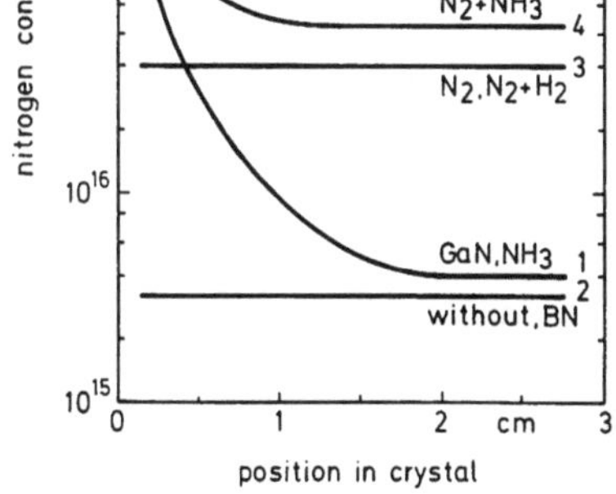

Fig. 9. Nitrogen doping profiles for different sources

method, however. The catalytic effect of Os is possibly suppressed in the presence of phosphorus vapour. Starting with mixtures of nitrogen and ammonia some improvement was achieved, especially in the region near the seed (curve 4). The best results were obtained using phosphorus nitride (P_3N_5) as a doping source[23]. P_3N_5 is a solid which is stable at temperatures below approximately 500 °C. At higher temperatures it decomposes producing active nitrogen. With a phosphorus nitride reservoir at a temperature of about 650 °C fixed between crucible and phosphorus reservoir the nitrogen doping is constant at approximately 2×10^{17} cm^{-3} (curve 5). This doping method can probably be further improved to yield GaP crystals with a homogeneous nitrogen concentration in the 10^{18} cm^{-3} range.

2.3 Other Binary Compounds

InP has for years been in the shadow of GaAs because of its slightly inferior properties and difficulty of growth. However, it is now emerging as an important substrate material in, for example, the epitaxial growth of lattice matched GaInAsP layers for use as double heterostructure lasers emitting in the 1.2 to 1.3 μm wavelength region[24] and heterojunction solar cells[25]. The TGS growth of InP single crystals was initiated by the authors[26]. Growth conditions are very similar to those of the GaP single crystals where the melt temperature was chosen to reveal a phosphorus concentration in the solution of approximately 5%[27, 28]. The properties of the crystals in[26] were found to be slightly inferior to those in crystals grown by the LEC method, but twin free material suitable for substrate applications was efficiently produced. Yamamoto and Uemara[29] recently reported that the method could be improved upon from a purity point of view by careful baking of the indium melt before growth. Ito and Ito[30] also grew InP recently by a similar method. Here, presynthesized InP polycrystals floating on the surface of the indium melt acted as the phosphorus source. Although the grown single crystals were reported to be somewhat smaller than those described in[26], controlled p-type doping with Zn and Cd was discussed.

With this growth method it should be possible to grow most of the III-V binaries, the exceptions being the compounds of boron and nitrogen. It will be recalled that one of the major advantages in the growth of the phosphides is that it takes place at normal pressure (1 bar) whereas growth from a stoichiometric melt requires an encapsulant and a high pressure chamber to compensate for the high dissociation pressures of these compounds at their melting points. However, this is not the case with the arsenides and antimonides where the binaries of these compounds exhibit intermediate (GaAs: 0.98 bar) and low (InSb: 10^{-7} bar) dissociation pressures at their melting points. Therefore, it is not known at the present time whether the TGS growth technique offers any distinct advantage over conventional growth methods in the production of these particular compounds.

Looking further afield, it should also be possible to grow a few of the *IV–VI compounds* by this method. Some of the constituent elements display basic requirements for this type of growth: a low vapour pressure for the melt component (e.g., Pb) and a high vapour pressure for the solute (e.g., S). Indeed, PbS has been grown[31]. The Pb-melt temperature was again chosen to reveal a sulphur concentration in the solution of approximately 5%. The material obtained was compact and polycrystalline, but was shown to contain a few Pb inclusions by electron microprobe analysis. Further experiments are, therefore, necessary to optimize growth conditions to omit inclusions and obtain directives for single crystal growth.

Rodot et al.[3] also mentioned the possibility of growing *II–VI compounds* by this technique. However, a further difficulty is introduced due to the fact that both groups of elements exhibit fairly high vapour pressures. An interesting application of a similar technique may be in the growth of *SiC* crystals as suggested by von Münch[32]. This particular material is of special interest because its favourable energy gap makes it a candidate for realization of blue LED's.

2.4 Ternary Compounds

The TGS growth technique has also been applied to various combinations of III-V ternary alloys. For example, Rodot et al.[3] mentioned the possibility of growing $GaAs_{1-x}P_x$ by this technique and presented a growth configuration. Homogeneous bulk crystals of this material grown by this method were indeed realized by Saito and Seki[33]. They termed the method solution growth by a modified travelling solvent technique. Polycrystals across the complete composition range were obtained by varying the temperature of an arsenic source between 500 °C (0.1 bar) and 570 °C (0.4 bar). The phosphorus source was again held at 420 °C to produce a 1 bar vapour pressure. The mixed vapours then reacted with the exposed surface of the gallium melt and growth followed in the manner described earlier in Sect. 2.1. Essential features for homogenous growth were reported to be that the temperature gradient in the gallium melt is not too large (diffusion of phosphorus slower than arsenic in gallium) and the group V sources be separated by a very thin capillary (prevents diffusion of As vapour into the colder P source region).

Although alloys of $GaAs_{1-x}P_x$ grown via the vapour phase dominate the commercial field of LED's at the present time, Archer[34] demonstrated a few years back that the theoretical quantum efficiencies of the alloys $In_{1-x}Al_xP$ and $In_{1-x}Ga_xP$ are almost a 100 times greater than $GaAs_{1-x}P_x$ for energies greater than 2 eV. These compounds would therefore seem to be superior materials for emissions in the yellow ($In_{1-x}Ga_xP$) and pure green ($In_{1-x}Al_xP$) parts of the visible spectrum, respectively. However, technological difficulties in material synthesis of these compounds have prevented them from becoming serious candidates for light source applications. For example, the growth of epitaxial layers of $In_{1-x}Al_xP$ from solution presents large problems due to the extremely high segregation coefficient of aluminium in indium. This effect was observed by Onton and Chicotka[35] who synthesized the alloy in bulk form by a TGS growth technique. They reported the compositions of ingots grown from the In-Al melt as varying between AlP and InP with the pure binaries themselves observed at the ends of the ingots. $In_{1-x}Ga_xP$ synthesis also presents difficulties such as a large segregation coefficient of Ga in In and the large lattice mismatch between the two binaries, but these can be overcome. The large separation of the solidus[36] and liquidus[37] curves in the temperature – composition diagram shown in Fig. 10 represents the large segregation of GaP for melt-

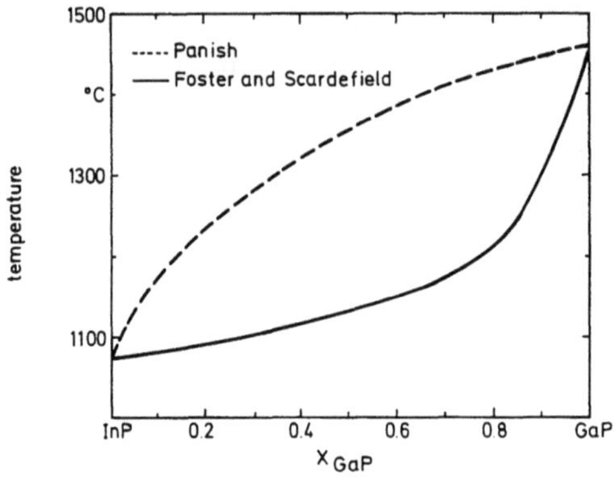

Fig. 10. Liquidus and solidus in the InP-GaP system

growth techniques, thus making the preparation of bulk alloys of uniform composition difficult by this method.

Prerequisites for the growth of homogeneous bulk $In_{1-x}Ga_xP$ alloys from solution are growth at constant temperature and a constant Ga content within the solution. Attempts to realize these requirements have been made by various authors: Synthesis of the material in small sealed crucibles at a constant temperature was effected by Burnham et al.[38)] and Macksey et al.[39)]. In each case In melts within a temperature gradient were supplied with source GaP to maintain a constant composition of the solution. However, Burnham et al.[38)] limited their paper to the discussion of only one composition, but indicated that the composition of the grown alloy was a direct function of the GaP to InP ratio in the source. Macksey et al.[39)] used crushed InP and GaP source material. Composition control was made by varying the growth temperature and selecting the In/Ga ratio of the solution which in turn could be regulated by the amount of source GaP or by adding Ga to the melt. However, the reproducibility of the particle size of a crushed source seems questionable. The area exposed for dissolution, which determines the rate of dissolution is therefore not well defined. Laugier and Chevallier[40)] compared two possibilities to grow bulk $In_{1-x}Ga_xP$. The first using an In-Ga melt continuously fed with P, as expected, produced relatively inhomogeneous crystals due to segregation effects. The second method they described was also adopted by Kato et al.[41)] and incorporated an In melt fed with source GaP set on the surface of the liquid and P from the vapour. Although both groups of workers[40, 41)] reported success in growing small quantities of homogeneous material across the complete solution range, details of composition control were omitted. In[40)] the two examples given were reported to be reasonably homogeneous. The composition was adjusted by varying the temperature, that is, partial pressure of the phosphorus source. Itoh et al.[42)] in a brief communication also discussed growth by a temperature gradient technique incorperating source GaP in an In melt, but within a horizontal growth system. Again growth of small amounts of homogeneous material was reported. More recently, Voigt et al.[43)] carried out investigations of growth defects of $In_{1-x}Ga_xP$ grown by the temperature gradient solution growth technique by means of a scanning electron microscope operating in the cathodoluminescent mode. The growth system was similar to earlier publications[40, 41)], but a little more detail of the actual growth mechanism was given.

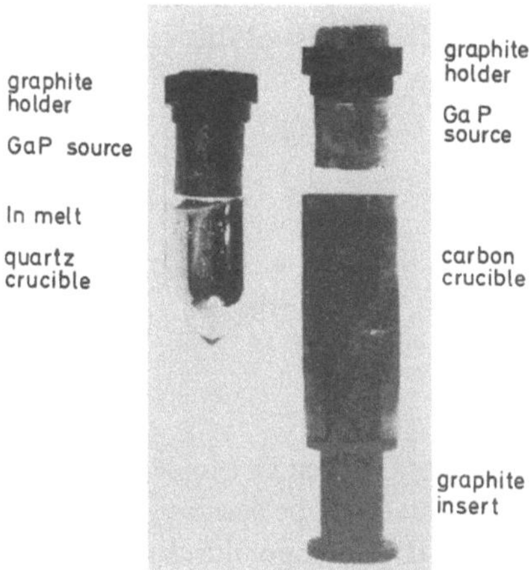

Fig. 11. Crucible configuration used for $In_{1-x}Ga_xP$ growth

However, although most authors report homogeneous material, none give exact details of how the desired composition could be determined prior to growth. This deficiency was removed by a variation of the TGS method[44]. The method enables the required crystal composition to be predetermined. The polycrystalline GaP source material shown in Fig. 11 is fixed in the graphite holders in a manner that when the system is "at temperature" its lower end is in contact with the surface of the indium melt. This then dissolves and due to the temperature gradient and concentration difference, gallium and phosphorus are transported to the lower colder end of the crucible where supersaturation causes crystallization. For clarity reasons the carbon crucible assembly is shown exploded. The graphite cylinder helps to control the heat flow at the onset of crystallization thus reducing the probability of inclusion formation. The size of the GaP source is so calculated that as gallium is removed from solution by crystallization, it is replenished at the rate necessary to keep the solution composition, and with it, the growing crystal composition constant. This is possible since the indium level climbs during the crystallization as little indium is used, and the volume of the crystal is larger than the volume of the liquid from which it is made. The process therefore takes advantage of a process that is normally detrimental to crystal growth, that is, a large segregation coefficient of one metal within another.

The crucible configuration is enclosed in an evacuated quartz ampoule at the lower end of which is red phosphorus at 420 °C. The phosphorus atmosphere at a pressure of 1 bar thereby prevents evaporation of phosphorus from the indium melt produced by the GaP source, and is used to form the InP content of the crystal. The ampoule is mechanically lowered at a rate equal to the growth rate to maintain constant growth-front conditions. In[44] details are to be found of how the size of the source is calculated and of thermodynamical considerations as to the process feasibility.

The growth of single crystals of $In_{1-x}Ga_xP$ by the TGS growth technique still remains a problem, although graded compositions from a GaP seed were reported[44]. The fact that the grain size of 1–3 mm did not vary appreciably over the ingot length is a directive that the growth conditions need to be controlled even more accurately. However, Miller and Baxter[45] also demonstrated the effectiveness of the method when growing GaSb rich boules of $In_{1-x}Ga_xSb$ with low defect concentrations at temperatures down to as low as 330 °C which indicates a good degree of flexibility.

3 Crystal Properties

3.1 Crystallographic Defects

The formation of *grain boundaries* during TGS growth of GaP can be avoided by seeding and careful control of growth conditions[7]. It can safely be assumed that this problem is also solvable for growth of other binary III-V compounds. In the case of InP it was demonstrated that this assumption is realistic[26]. With ternary III-V compounds the problem is much more difficult. First attempts with $In_{1-x}Ga_xP$ did not result in growth of single crystalline material[44]. The reason for the increased difficulties in growing ternary compounds is that their phase diagrams are more complicated. Especially the formation of cellular structures and striations can hardly be avoided[43].

Crystals grown by the TGS method have relatively low *dislocation densities.* The values for GaP are typically in the range 5×10^4 to 5×10^5 cm^{-2} [6, 7, 46]. The same order of magnitude was also found with InP[26]. On the other hand these values re-

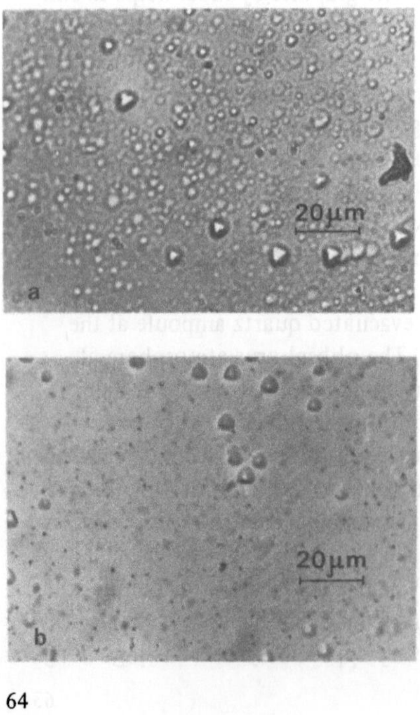

Fig. 12a and b. Etch pits on GaP crystal surfaces. **a** LEC grown; **b** TGS grown[7]

present also the state of the art for LEC grown crystals. Therefore, the TGS growth methods does not offer an advantage in this respect. However, the dislocations in TGS grown crystals seem to be less decorated with impurity atoms due to the lower growth temperature. This difference was observed by Tajima and Iizuka[46] using a light scattering method.

Another type of defect can be detected by etching studies. From their shape they are called *S pits* ("saucer") and they can easily be distinguished from the normal D pits ("dislocation"). For a long time they were considered to be due to impurity precipitates and/or vacancy clusters[47, 48]. However, recent studies[49] have shown that these defects are more likely small dislocation loops. Anyhow, the occurrence of the S pits is closely correlated to the luminescence efficiency[50]. The S pit density in GaP LEC crystals is usually around 10^6 cm^{-2} [5, 6], whereas TGS crystals exhibit less than 10^4 cm^{-2} S pits[5–7]. Figure 12 shows this difference clearly.

Antisite defects are also discussed as possible "killer centres" in GaP luminescent material[51]. These defects can be detected by electron spin resonance in LEC crystals, in TGS material they do not, however, appear[52].

3.2 Impurities and Electrical Properties

Only few data on the incorporation of impurities in TGS grown crystals are available. Undoped GaP crystals[4] were p conducting with an acceptor concentration of 1.4×10^{17} cm^{-3}. The acceptor species was identified by mass spectrometry as silicon from the quartz crucible. Similar results were found in[6]: p around 10^{17} cm^{-3} with silicon as the acceptor. In Ref.[5] silicon from the quartz crucible acts as donor, resulting in carrier concentrations slightly above 10^{17} cm^{-3}. More recent studies[7, 8] have shown that n-type conduction is typical for undoped TGS grown crystals. Carrier concentrations below 10^{15} cm^{-3} also are possible[7, 8]. The use of carbon crucibles in Ref.[7] caused some carbon contamination (120 to 220 ppm). However, most of this acceptor-type impurity is electrically inactive because the crystals are n conducting. In Table 1 carrier concentrations and mobilities in TGS grown GaP crystals are summarized. The mobility values show that the material is usually compensated.

Table 1. Electrical properties of undoped TGS-GaP

	Carrier concentration (cm^{-3})	Mobility (cm^2/Vs)	Ref.
n (300 K)	6×10^{14}	200	8)
	1×10^{15}	75	7)
	7×10^{16}	83	18)
n (77 K)	1×10^{13}	420–1000	3)
p (300 K)	6.6×10^{16}	100	6)
	8.3×10^{16}	115	4)
p (77 K)	8×10^{14}	455	6)

GaP crystals can be doped n-type with sulphur, selenium or tellurium (see Sect. 2.2.4). The doping dependence of carrier concentration and mobility was studied thoroughly in the case of S and Te[5, 6, 18] and was found to be similar to the behaviour of LEC grown crystals.

Carrier concentration and mobility of undoped InP crystals were measured to be $n = (5-10) \times 10^{16}\ cm^{-3}$ and $\mu_n = 3700\ cm^2/Vs$, respectively[26], which is again very similar to the corresponding values in LEC material.

4 Devices

4.1 Red and Green LED's from GaP

The most important application of GaP crystals is their use as a semiconductor material for light emitting diodes. As GaP is an indirect semiconductor, the luminescent efficiency of undoped GaP is very low. However, by introducing isoelectronic centres efficient light emission can be achieved. Zinc-oxygen pairs give rise to red emission (peak wavelength 690 nm at room temperature) and nitrogen centres emit green light (560 nm). Both types of GaP LED's have been realized with crystals grown by the TGS (or SSD) growth technique. Red LED's have been produced by single liquid phase epitaxy of Zn-O doped GaP on SSD grown substrates[5, 8]. The quantum efficiencies stated (up to 10% and 4–5%, respectively) are approximately equal to the state of the art of double liquid phase epitaxy on LEC substrates. The degradation of these devices depends on the donor concentration in the n region[53]. In Ref.[5] green LED's with a quantum efficiency of 0.05% are also mentioned. These were fabricated by double liquid phase epitaxy. Therefore the SSD grown substrate

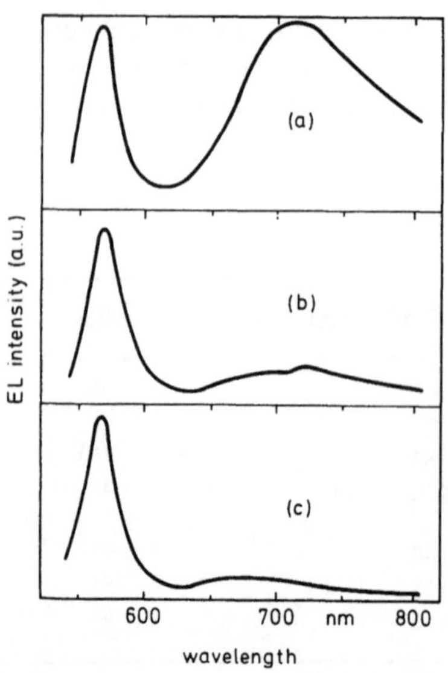

Fig. 13. Electroluminescence (EL) spectra of bulk GaP LED's[22]

is not a part of the active region of these devices. An increase of the red emission component was observed during degradation of the diodes[54].

Green LED's from bulk GaP grown by the TGS method were first described by Poiblaud and Jacob[6]. The p-n junction was produced by zinc diffusion into sulphur doped material. However, because the crystals were not doped with nitrogen, the quantum efficiency of these devices was only 0.0007%. In Ref.[22] green LED's were also fabricated by zinc diffusion into sulphur doped GaP crystals. Here the material was doped with nitrogen up to 3×10^{17} cm^{-3}. Consequently much higher efficiencies (0.01%) were obtained, which are comparable to those of commercially available vapour phase epitaxial diodes. Room temperature electroluminescence spectra of three LED's from bulk SSD-GaP are shown in Fig. 13[22]. Besides the green emission at 565 nm these spectra exhibit a broad red peak around 700 nm which is related to residual oxygen. In Fig. 13b it is indicated that the 700 nm peak consists of two parts: One at 690 nm, which is due to Zn-O pairs on the p side of the junction, and one at 720 nm, the origin of which is not yet clear. The first diode (a) was fabricated from a crystal growth run taking no special precautions with respect to oxygen contamination. The crystal from which the second diode (b) was made was grown from a carefully outgassed melt. In the third case (c), the melt was not only outgassed, but the ampoule contained red phosphorus completely free of white phosphorus, thereby carrying less oxygen in the form of P_2O_5 into the system. Here only the 690 nm peak is present with low intensity.

Further development work is necessary to make bulk GaP green LED's fabricated from TGS grown crystals fully competitive with LED's produced using vapour or liquid phase epitaxial material. A tenfold increase in nitrogen doping concentration seems feasible, which will bring the quantum efficiency into the 0.1% range. This value is also achieved today with top quality vapour phase epitaxial diodes. The crystal diameter should be increased to at least 50 mm, to make device production from this material economically comparable to standard production processes.

4.2 Red to Yellow LED's from $In_{1-x}Ga_xP$

Although it is theoretically possible to obtain devices emitting from the infrared to the yellow in the $In_{1-x}Ga_xP$ ternary system, we will restrict ourselves in this section to electroluminescent emission from the red to the yellow parts of the visible spectrum reported in this alloy grown by the temperature gradient solution growth technique. It will be recalled that it is in this region that the theoretical quantum efficiencies calculated by Archer[34] start to show any distinct improvement over similar systems such as $GaAs_{1-x}P_x$.

Lorenz et al.[55] reported electroluminescent emission from alloys grown by this technique. However, as segregation effects were not considered during growth, the Te-doped ingots revealed a decrease in the Ga/In ratio in the growth direction seen as a random variation in the half-width of emission lines of Zn diffused p-n junctions. The maximum direct transition energy gap was reported to occur at x = 0.8 with an energy of 2.25 eV at 300 K. The quantum efficiency, described as being independent of energy, varied between 1 and 3×10^{-4} increasing a 1000 fold at 77 K.

Macksey et al.[39] grew N-doped alloys and associated the recombination process in the Zn diffused diodes made from these crystals to Te-Zn donor-acceptor pairs which emitted at 50 meV lower energy at 77 K than the photoluminescent peak measured associated with NN pairs. Kato et al.[56] associated the recombination process of Zn diffused devices to conduction band-Zn acceptor transitions which they confirmed by a comparison of electroluminescence and photoluminescence measurements. However, the composition described was x = 0.35, just out of the range of interest here. Electroluminescent emission in the spectral region red to yellow in this alloy system was reported in[57]. The ingots grown for this purpose[44] were extrinsically doped n-type by addition of a small amount of elemental sulphur to the growth ampoule. The quantity of sulphur was the same as that described in Sect. 2.2.4 on doping during the growth of GaP single crystals by this method where a carrier concentration of $5 \times 10^{16} - 2 \times 10^{17}$ cm^{-3} was obtained[7]. Indeed, values of carrier concentration varied between samples from typically 5×10^{16} to 4×10^{17} cm^{-3} at 300 K with no definite dependence on alloy composition being observed. These values decreased only slightly at 77 K to typically 3×10^{16} cm^{-3} compared to 5×10^{16} cm^{-3} at 300 K.

From polished slices of polycrystalline $In_{1-x}Ga_xP$ with compositions ranging between x = 0.5 to 0.8 p-n junctions were fabricated using a mixed Zn-Ga diffusion source in closed quartz ampoules at 860 °C for one hour. With GaP comparison samples these diffusion conditions yielded a diffusion depth of 5 μm. In the polycrystalline $In_{1-x}Ga_xP$ samples the diffusion depth was difficult to determine, but it is most probably larger than 5 μm as to be expected from the behaviour of other mixed crystals. After diffusion the backside p-layer was removed by lapping and polishing. Ohmic contacts were applied to the p-side by Au-3% Zn dots, whereas

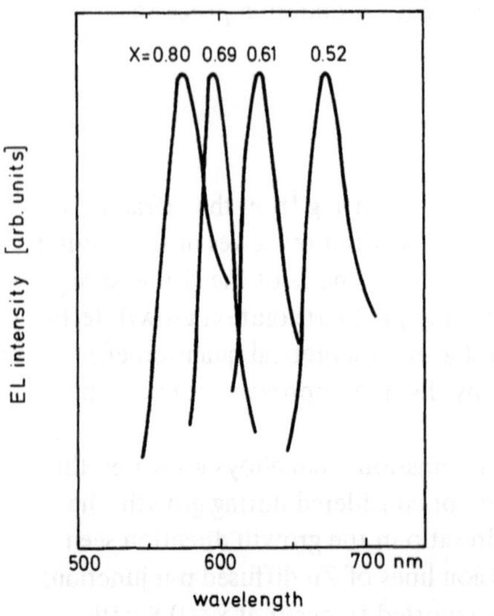

Fig. 14. Electroluminescence (EL) spectra of $In_{1-x}Ga_xP$ LED's

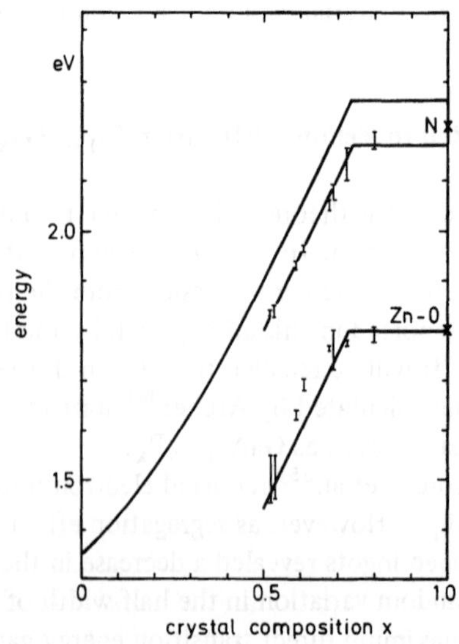

Fig. 15. Emission peak energy of $In_{1-x}Ga_xP$ LED's

the n-side was covered with a full area Au-3% Ge contact. Diode separation was afforded by a diamond saw resulting in chips of 450 μm side length. Electroluminescence spectra and quantum efficiency were measured using standard techniques at a diode current of 25 A cm^{-2}. Figure 14 shows normalised spectra of four diodes fabricated from material of different compositions. The emission peak is shifted from 675 nm, corresponding to a dark red, to 570 nm (greenish yellow) with increasing gallium content. The peak width of 30 nm is about the same as that found in GaP LED's and is a further proof of material homogeniety. In Fig. 15, the emission peak energy of the LED's is compared with the band gap of $In_{1-x}Ga_xP$ as a function of composition after Stringfellow et al.[58] given by the solid line. The energetic difference of approximately 90 meV can be explained by conduction band – acceptor state transitions in accordance with earlier publications[59]. In addition to this near band gap emission a broad peak of varying intensity at lower energies was always present. The peak energy of this second emission was around 0.45 eV below the band gap. As this energy difference is also found in Zn-O doped red emitting GaP diodes, the lower energy peak is also probably due to Zn-O pairs.

In Fig. 16 the results of quantum efficiency measurements of the $In_{1-x}Ga_xP$ diodes are shown. The efficiency is apparently strongly dependent on the crystal composition. In the direct range typical efficiencies were 7 x 10^{-5} with a maximum of 1.5 x 10^{-4}. Near the direct-indirect crossover point the efficiency drops drastically with values of 10^{-5} in the indirect range. There are two explanations for the rather low efficiencies, (a) the material was polycrystalline and (b) the p-layer thickness was not optimized. The latter explanation is supported by the observation that in some cases the side faces of the diodes were much brighter than the top area.

Improvement of diode performance is in this case a direct function of crystal quality. If single crystals of this compound are available, the situation regarding electroluminescent emission from $In_{1-x}Ga_xP$ diodes can once again be reviewed.

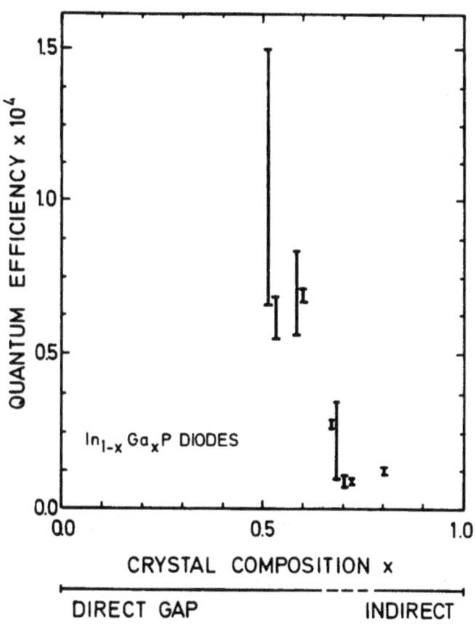

Fig. 16. Quantum efficiency of $In_{1-x}Ga_xP$ LED's

5 References

1. Thurmond, C. D.: J. Phys. Chem. Solids *26*, 785 (1965)
2. Jordan, A. S., Von Neida, A. R., Caruso, R., Kim, C. K.: J. Electrochem. Soc. *121*, 153 (1974)
3. Rodot, H., Hruby, A., Schneider, M.: J. Crystal Growth *3/4*, 305 (1968)
4. Besselere, J. P., Le Duc, J. M.: Mat. Res. Bull. *3*, 797 (1968)
5. Kaneko, K. et al.: Proc. IEEE *61*, 884 (1973)
6. Poiblaud, G., Jacob, G.: Mat. Res. Bull. *8*, 845 (1973)
7. Gillessen, K., Marshall, A. J.: J. Crystal Growth *32*, 216 (1976)
8. Moravec, F., Novotný, J.: J. Crystal Growth *33*, 90 (1976)
9. Broder, J. D., Wolff, G. A.: J. Electrochem. Soc. *110*, 1150 (1963)
10. Plaskett, T. S., Blum, S. E., Foster, L. M.: J. Electrochem. Soc. *114*, 1303 (1967)
11. Blum, S. E., Chicotka, R. J.: J. Electrochem. Soc. *115*, 298 (1968)
12. Blum, S. E., Chicotka, R. J., Bischoff, B. K.: J. Electrochem. Soc. *115*, 324 (1968)
13. Blum, S. E., Chicotka, R. J.: J. Electrochem. Soc. *120*, 588 (1973)
14. Woodbury, H. H.: J. Crystal Growth *35*, 49 (1976)
15. Von Neida, A. R., Oster, L. J., Nielsen, J. W.: J. Crystal Growth *13/14*, 647 (1972)
16. Frosch, C. J., Derick, L.: J. Electrochem. Soc. *108*, 251 (1961)
17. Gillessen, K.,Marshall, A. J.: unpublished
18. Menninger, H. et al.: Phys. Stat. Sol. (a) *48*, 407 (1978)
19. Gillessen, K., Marshall, A. J.: J. Crystal Growth *33*, 356 (1976)
20. Scheel, H. J.: J. Crystal Growth *13/14*, 560 (1972)
21. Kaneko, K. et al.: J. Electrochem. Soc. *121*, 556 (1974)
22. Gillessen, K., Marshall, A. J., Schuller, K.-H., Gramann, W.: IEEE Trans. Electron Devices *ED-24*, 944 (1977)
23. Poiblaud, G.: French Pat. 7 431 524 (1974)
24. Hsieh, J. J., Shen, C. C.: Appl. Phys. Lett. *30*, 429 (1977)
25. Wagner, S. et al.: J. Crystal Growth *39*, 128 (1977)
26. Marshall, A. J., Gillessen, K.: J. Crystal Growth *44*, 651 (1978)
27. Panish, M. B.: J. Crystal Growth *27*, 6 (1974)
28. Sol, N., Clariou, J. P., Linh, N. T., Moulin, M.: J. Crystal Growth *27*, 325 (1974)
29. Yamamoto, A., Uemara, C.: Jap. J. Appl. Phys. *17*, 1869 (1978)
30. Ito, K., Ito, H.: J. Crystal Growth *45*, 248 (1978)
31. Marshall, A. J., Gillessen, K., Daniel, D. R.: unpublished results
32. Von Münch, W.: private communication
33. Saito, T., Seki, Y.: J. Crystal Growth *23*, 217 (1974)
34. Archer, R. J.: J. Electr. Materials *1*, 128 (1972)
35. Onton, A., Chicotka, R. J.: J. Appl. Phys. *41*, 4205 (1970)
36. Foster, L. M., Scardefield, J. E.: J. Electrochem. Soc. *117*, 534 (1970)
37. Panish, M. B.: J. Chem. Thermodynamics *2*, 1301 (1970)
38. Burnham, R. D. et al.: Appl. Phys. Lett. *17*, 430 (1970)
39. Macksey, H. M. et al.: J. Appl. Phys. *44*, 1333 (1973)
40. Laugier, A., Chevallier, J.: Phys. Stat. Sol. (a) *7*, 427 (1971)
41. Kato, T., Shimizu, A., Ishida, T.: Jap. J. Appl. Phys. *13*, 1481 (1974)
42. Itoh, H., Hara, K., Tanaka, A., Sukeyaura, T.: Appl. Phys. Lett. *19*, 348 (1971)
43. Voigt, G. et al.: Phys. Stat. Sol. (a) *36*, 173 (1976)
44. Marshall, A. J., Gillessen, K.: J. Crystal Growth *41*, 93 (1977)
45. Miller, J. F., Baxter, R. D.: J. Electronic Materials *2*, 485 (1973)
46. Tajima, M., Iizuka, T.: Jap. J. Appl. Phys. *15*, 651 (1976)
47. Iizuka, T.: J. Electrochem. Soc. *118*, 1190 (1971)
48. Rozgonyi, G. A., von Neida, A. R., Iizuka, T., Haszko, S. E.: J. Appl. Phys. *43*, 3141 (1972)
49. Werkhoven, C.: J. Electrochem. Soc. *125*, 671 (1978)
50. Rozgonyi, G. A., Afromowitz, M. A.: Appl. Phys. Lett. *19*, 153 (1971)
51. Van Vechten, J. A.: J. Electronic Materials *4*, 1159 (1975)

52. Kaufmann, U., Schneider, J., Räuber, A.: Appl. Phys. Lett. *29*, 312 (1976)
53. Isawa, N., Kaneko, K.: Jap. J. Appl. Phys. *18*, 691 (1979)
54. Kaneko, K.: Jap. J. Appl. Phys. *15*, 1287 (1976)
55. Lorenz, M. R. et al.: Appl. Phys. Lett. *13*, 421 (1968)
56. Kato, T., Matsumoto, T., Ishida, T.: Jap. J. Appl. Phys. *15*, 2265 (1976)
57. Gillessen, K., Marshall, A. J.: 7th ESSDERC, Brighton 1977
58. Stringfellow, G. B., Lindquist, P. F., Burmeister, R. A.: J. Electronic Materials *1*, 4 (1972)
59. Onton, A., Lorenz, M. R., Reuter, W.: J. Appl. Phys. *42*, 3420 (1971)

Received June 25, 1979

Molecular Beam Epitaxy of III-V Compounds

Klaus Ploog

Max-Planck-Institut für Festkörperforschung, D-7000 Stuttgart 80, Federal Republic of Germany

An important aspect of the large expansion in the development and production of solid-state devices has been the increasing demand for more sophisticated multilayered semiconductor crystals. During the last decade a new thin film growth technique, molecular beam epitaxy (MBE), has been developed to provide improved control over composition, thickness, and doping profile in the direction of growth on an atomic scale. Based on these unique features, molecular beam epitaxy offers the feasibility of tailoring the electronic properties of the crystal to its desired function by growing high-quality epitaxial thin films with predetermined compositional and/or doping profiles perpendicular to the growth surface. The attractive capabilities of MBE for large-area uniform growth with extreme control over microscopic dimensions suggest that both characteristic and reproducibility of microwave and optoelectronic device structures may be improved. It is the purpose of this article to review the basic process of molecular beam epitaxy as a crystal growth technique, to indicate present trends and material problems, and – at a time of continuing innovation – to show promise for future application. The emphasis of this review is on the growth and properties of crystalline III-V compound semiconductor thin films focusing mainly on GaAs and $Al_xGa_{1-x}As$.

Table of Contents

1 Introduction

The method of molecular beam epitaxy (MBE) is a relatively new epitaxial growth technique in a conceptual and exploratory sense compared to the commonly used techniques of liquid phase epitaxy (LPE) and vapour phase epitaxy (VPE). Therefore, it will be useful to define what is meant by this phrase in the present context. According to a definition given by Cho and Arthur[1], who re-investigated the fundamental process, the term molecular beam epitaxy describes the epitaxial growth of compound semiconductor films by the reaction of thermal-energy molecular beams of the constituent elements with a crystalline substrate surface held at a suitable temperature under ultra-high vacuum (UHV) conditions. Thus UHV techniques are required throughout, and a MBE system is basically a vacuum evaporation apparatus.

It was this crucial requirement of improved vacuum conditions for successful epitaxial growth which had retarded a rapid development of the original ideas of MBE, first established by Günther[2] in 1958, until the middle 1960's when commercial ultra-high vacuum systems became more readily available. Using multiple molecular beams, Günther was the first to describe in general the growth of III–V materials. His "three-temperature" technique was based on the much higher vapour pressure of the group V element over the pure group V solid than over the III–V compound. Günther was able to produce stoichiometric films of both binary and ternary III–V compounds using a group V element source oven kept at temperature T_1 to maintain a steady pressure in a static vacuum chamber. The source oven with a group III element at a higher temperature T_3 was providing a flux of atoms incident on the substrate, which was critical for optimum condensation rate. The substrate was held at an intermediate temperature T_2, which had to be high enough to eliminate the condensation of excess group V element on the surface. However, Günther's films were not epitaxial, but grown on glass substrates and hence were polycrystalline[2].

It was not until 10 years later that epitaxy for monocrystalline GaAs was achieved by Davey and Pankey[3] using Günther's method. They employed clean single crystal substrates under highly improved vacuum conditions. At the same time, Arthur[4] conducted fundamental studies of the kinetic behaviour of Ga and As_2 species on GaAs surfaces that led to a first rough understanding of the growth mechanism. Subsequently, epitaxial thin films of GaAs and related III–V compounds of superior quality and extreme smoothness have been grown by the MBE technique by Arthur, Cho, Chang, and coworkers[1, 5]. Recently the rate of development and innovation in the field of molecular beam epitaxy and the application of the technique to other materials, including II–VI and IV–VI compounds, has been very rapid. Using MBE it is now possible to produce ultra-thin layer structures consisting of as many as 10^4 alternate monolayers of GaAs and AlAs[6], which might be the prototype of synthesizing a new class of alloys. The most significant advances in device fabrication, where MBE has made essential contributions, include III–V compound microwave devices (FET's; Varactor, IMPATT, and Mixer diodes) and CW injection lasers, optical waveguides, and optical couplers for integrated optics-systems[7]. In addition, MBE of IV–VI compounds has been improved significantly to the point of fabrication of double-heterostructure lasers with CW operations[8].

The object of this review is to outline the basic technology of the MBE process as a film growth technique and to give some brief perspectives of the research applications and of the device capabilities of films prepared by MBE. The discussion will be limited to films of the III–V compounds and will be mainly focussed on GaAs and $Al_xGa_{1-x}As$, the MBE material from which most of the actual devices have been fabricated. Our understanding of the growth and doping mechanisms, while still incomplete, is probably best for this material system.

The review will begin with a discussion of some general features inherent to the MBE process. In what follows, the equipment and some practical experimental details will be described at some length because of the considerable difference between MBE and LPE or VPE. In Sect. 2.3 the growth mechanism and dopant incorporation will be discussed as it is presently understood. Electrical and optical properties of MBE grown GaAs films will be presented in Sect. 3. These parameters may help place the material in perspective with respect to the films grown by LPE and VPE. Finally, a brief indication of some of the device capabilities of films produced by MBE and an illustration of some applications of MBE in fundamental studies will conclude this review. Particular emphasis will be put on recent developments and advances since about 1975. The historical background of MBE and its development up to the end of 1974 have been the subject of an excellent and comprehensive review by Cho and Arthur[1)]. Recently, another article on this subject with particular emphasis on kinetic aspects has been written by Joyce and Foxon[9)].

Most of the principles and techniques to be described extend to other material systems with little or no modifications. The growth and application of both II–VI and IV–VI compound epitaxial films by MBE has been thoroughly reviewed by Chang and Ludeke[10)], Farrow[11)], and most recently by Smith[12, 13)].

2 Crystal Growth

2.1 General Considerations

Intensive research activities in the last five years have developed the MBE growth technique to a stage where it has now reached a pronounced technological interest. Compared to the more established methods of thin film growth the most distinguishing characteristics of MBE are

(i) a low growth rate (between 1 and 3 Å/s for GaAs);
(ii) a low growth temperature ($T_s \leqslant 630$ °C for GaAs);
(iii) the ability of abrupt cessation or initiation of growth;
(vi) a smoothing of the surface of the growing crystal during growth down to atomic steps.
(v) the facility for in situ analysis.

As a result of these favorable features the MBE technique provides a reproducible control over composition, thickness, and doping profile in the direction of growth on an atomic scale, and it has already demonstrated an excellent uniformity of crystal properties over the whole area of the substrate. To the production of semiconduc-

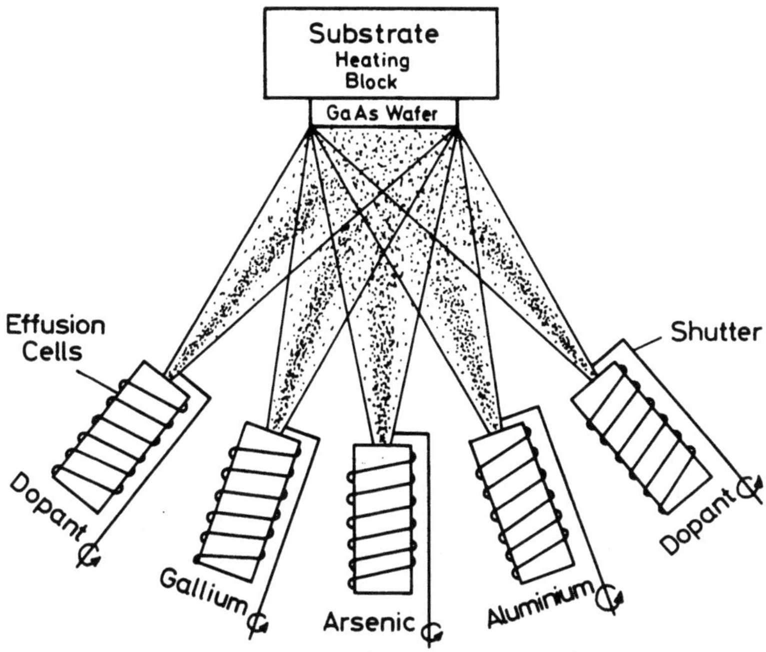

Fig. 1. Schematic illustration of the essential elements for molecular beam epitaxy of doped gallium arsenide and aluminium gallium arsenide

tor heterostructures and doping profiles this contributes approximately one order of magnitude improvement with respect to performance resolution in the direction of growth beyond the more conventional techniques.

With molecular beam epitaxy, the epitaxial films grow from thermal beams of atoms or molecules ("molecular beams") of the constituent elements impinging onto a heated crystalline substrate. The basic process is schematically illustrated in Fig. 1, which shows the essential elements for MBE of doped $Al_xGa_{1-x}As$. The molecular beams are generated from sources of the Knudsen-effusion-cell type[14)], whose temperatures are accurately controlled to $\pm 1\ ^\circ C$, in an ultrahigh vacuum system. Each source is provided with an externally controlled individual shutter. Operation of these shutters permits rapid changing of the beam species in order to abruptly alter the composition and/or doping of the growing film. With respect to the low growth rate of less than 1 μm/h, the shutter time is much less than the time for the growth of a monolayer. Thus, abrupt interfaces can be realized, if diffusion and surface accumulation are negligible at the low epitaxial growth temperature for MBE.

The intensities of the beams incident on the substrate crystal of temperature T_s are controlled by the temperatures of the effusion cells. Provided that the cell aperture is less than the mean free path of vapour molecules within the cell (that is Knudsen effusion), the flux F is given by[14, 15, 16)]

$$F_i = \frac{a \cdot p_i}{\pi \cdot d^2 \cdot \sqrt{2 \pi m_i \cdot k \cdot T}} \cdot \cos\theta \ [\text{molecules/cm}^2 \cdot \text{s}] \qquad (1)$$

where F_i is the flux of atoms or molecules of species i per unit area at the substrate, a is the area of the cell aperture in cm^2, p_i is the equilibrium vapor pressure of species i (in Torr) in the cell at the absolute temperature T, d is the distance of the orifice from the substrate in cm, m_i is the mass of the effusing species, and θ is the angle between the beam and the normal of the substrate. The group III element flux is always atomic, while the group V element beam can be either diatomic or tetratomic, depending on the source material. For typical growth rates of $\leqslant 1\ \mu m/h$ the fluxes used are approximately $10^{15}-10^{16}$ molecules/$cm^2 \cdot s$ for group V elements, $10^{14}-10^{15}$ atoms/$cm^2 \cdot s$ for group III elements, and 10^8-10^{11} atoms/$cm^2 \cdot s$ for dopants. They are produced with cell pressures of $10^{-3}-10^{-2}$ Torr for the main constituent elements and correspondingly lower pressures for dopant elements.

By choosing appropriate cell and substrate temperatures, epitaxial films of the desired chemical composition can be obtained. According to the phase diagram in Fig. 2, there is only a single compound formation in the Ga-As system corresponding to the composition GaAs. The deviation from stoichiometry is very small, $\delta y \leqslant 10^{-4}$. In Fig. 2, the deviation from stoichiometry has been greatly exaggerated to show its existence. The guidelines for achieving stoichiometric III–V compounds by MBE have been found in the surface chemical dependence of the sticking coefficient of the volatile group V element[4, 9, 17–19]. The parameter known as the sticking coefficient, S, describes the overall process of adsorption, diffusion, and growth. For each species S is defined as the fraction of the total impinging atoms or molecules that sticks to the surface and is incorporated into the film. At temperatures where epitaxial growth occurs, generally only that amount of the group V element adheres to the growing surface which collides with excess group III atoms on the surface and reacts to form the III–V crystal. As a result, one group V atom sticks for each group III atom and epitaxial growth on the III–V substrate occurs. Therefore, the growth rates of GaAs and $Al_xGa_{1-x}As$ are essentially determined by the Ga or Ga plus Al arrival rates, and their stoichiometry is ensured as long as an excess flux of arsenic molecules is supplied. For source material one can thus use either the compound

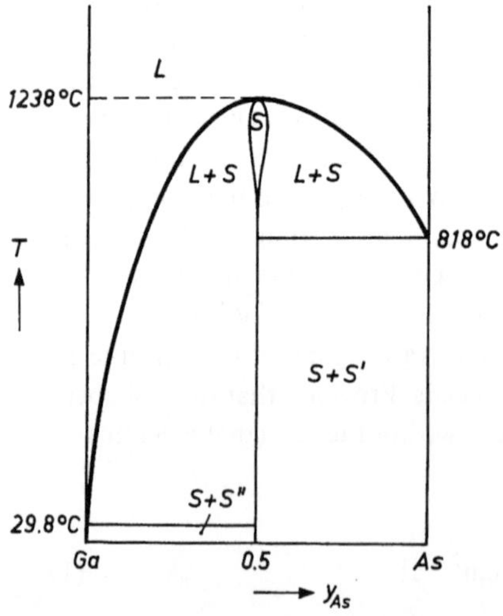

Fig. 2. Phase diagram of the system gallium-arsenic. There is only a single compound formation corresponding to the composition GaAs. The deviation from stoichiometry, which is very small ($\delta y \leqslant 10^{-4}$), has been greatly exaggerated in this schematic diagram in order to show its existence

itself, or its constituent elements, or their combination, and one is thereby able to control very accurately the growth rate and the surface environment in terms of the relative arrival rates. While this is a simplified interpretation of what is in reality a complex process (for details see Sect. 2.3), the basic validity of this growth model has made it possible to readily attain film stoichiometry and to grow high quality GaAs and $Al_xGa_{1-x}As$ crystals. The basic relationship between the sticking coefficients makes this system easy, convenient, and flexible.

The purity and thus the quality of the material to be deposited by molecular beam epitaxy is directly related to the extent of the reaction of the growing surface with the background gaseous species. The need for a growth system with improved vacuum conditions ($p \leq 10^{-9}$ Torr) arises from the low growth rate during MBE growth coupled with the stringent requirement of low unintentional impurity levels in the grown films. If background gaseous molecules with unity sticking coefficients were present in the reaction chamber at a pressure of $\sim 10^{-9}$ Torr, more than half of the film grown at a rate of ≤ 1 μm/h would consist of contaminants. Therefore, the partial pressures of gases with high sticking coefficient should be maintained below 10^{-14} Torr in the vicinity of the growing surface. This requirement is vital, for example, for the growth of high-quality $Al_xGa_{1-x}As$ where the growing surface is highly reactive with respect to CO and H_2O background species[7, 20]. For less reactive background species, the acceptable partial pressures can be slightly higher. In practice, during growth, the total pressure in the reaction chamber may rise above 10^{-9} Torr due to scattered beam species and, in the case of GaAs and $Al_xGa_{1-x}As$, due to the excess arsenic molecules offered. However, the UHV condition is always maintained with respect to impurity species.

While the procedure for attaining UHV conditions in the reaction chamber is time consuming, this environment has turned out to be ideal for surface analytical techniques which provide in situ examination of the substrate surface prior to epitaxial growth and a high degree of in situ growth control. The incorporation of a number of surface diagnostic techniques directly into the growth apparatus has contributed much to the progress which has been made in defining and understanding the general growth mechanism for MBE.

2.2 Technology of Molecular Beam Epitaxy

2.2.1 Basic MBE Growth Apparatus

The basic elements of a modern MBE growth system are
- UHV chamber
- substrate heater block
- individually shuttered and liquid nitrogen shrouded molecular beam effusion cells
- substrate exchange load-lock system.

In addition, the components of several appropriate analytical techniques are often attached directly to the MBE growth chamber. The associated instrumentation can provide essential facilities to monitor the residual gas and the molecular beam species, to control the surface structure of the growing film, or to detect any contamination of the substrate surface as well as to determine the surface composition of the grown epitaxial film. The analytical facilities have been continuously applied in-situ during molecular beam epitaxy, in order to ensure that the desired reaction conditions have been reached before and are being maintained during the crystal growth[20].

Until recently, most of the UHV apparatus, in which MBE growth experiments have been performed, were designed and built with instructions of the individual operators. Therefore, at the present time, the majority of the MBE systems used in various laboratories differ widely in both complexity and automation.

The system presently used by the author for MBE growth of GaAs and $Al_xGa_{1-x}As$ is schematically shown in Fig. 3. The system is of the vertical evaporation type and it was designed for a growth process providing abrupt changing of dopants (p to n-type) and of composition (GaAs to $Al_xGa_{1-x}As$) without interruption of growth. It illustrates the significant features of most MBE systems.

Five effusion cells are installed in the growth chamber containing elemental arsenic, gallium, aluminium and two appropriate dopant elements, for example Be and Si or Ge for p- and n-type material, respectively. The cells are thermally isolated from each other by several layers of corrugated tantalum foil wrapped directly around the outer alumina tube, and by tantalum radiation shields between adjacent cells. Each cell has its own shutter. The complete multiple effusion cell assembly is surrounded by liquid-nitrogen-cooled shrouds for cryopumping of condensible species in order to minimize outgassing into the reaction chamber. The temperatures of the

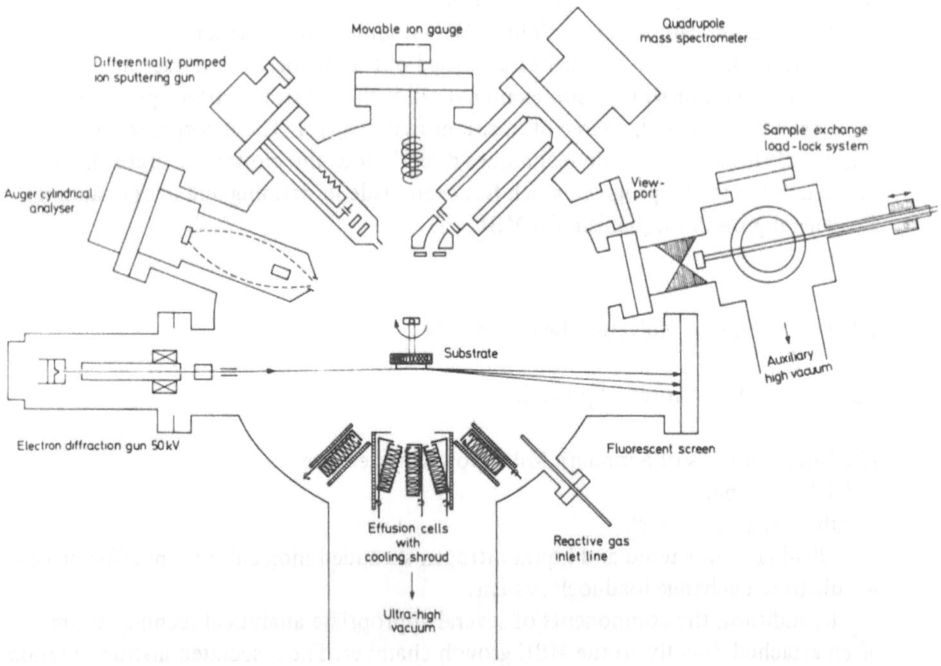

Fig. 3. Schematic cross-section of a typical UHV system designed for molecular beam epitaxy of doped GaAs and $Al_xGa_{1-x}As$ (vertical evaporation type)

individual effusion cells and of the substrate are measured with tungsten-rhenium (3% Re versus 25% Re) thermocouples, and they are monitored by a computer controlled temperature regulation system[21]. A reactive gas inlet line provides the controlled admission of gases including oxygen, nitrogen, ammonia, or water vapour, which are used either as dopants or for the deposition of nitrides and of insulating oxide films.

The substrate is mounted on a molybdenum heating block attached to a standard manipulator for precise positioning within the UHV chamber. The distance from the orifices of the cells to the substrate surface during MBE growth is about 70 mm. The substrate is loaded via a sample exchange load-lock system, which permits rapid introduction of single substrates while the pressure in the growth chamber is maintained in the 10^{-9} Torr range. The independently UHV-pumped loading chamber is separated from the main vacuum system by an all-metal valve. It is also used for substrate preparation prior to growth and for film structure processing after growth.

The UHV system indicated in Fig. 3 is equipped with a 400 l/s ion pump, a titanium sublimation pump with cryopanel, and a large diameter gate valve between growth chamber and pumping module. In various laboratories, alternative pumping systems including liquid-nitrogen-trapped diffusion pumps[9, 22], liquid-nitrogen-trapped turbomolecular pumps[23], and closed-loop helium cryopumps[24] have been employed in order to obtain clean UHV conditions in the MBE growth chamber. In the author's laboratory a combination of turbomolecular pump and titanium sublimation pump with cryopanel has been extensively used to deposit insulating oxide films onto MBE grown GaAs in the same growth run for application in GaAs MOS structures[23, 25]. A replacement of the ion pumps by a diffusion or turbomolecular pump (or a combination of both) is highly recommended, if large amounts of highly volatile elements like phosphorus (for example for deposition of $In_xGa_{1-x}P$[26]) are evaporated in the system.

In the horizontal bell-jar of 450 mm diameter used by the author, the molecular beam sources, the substrate, and the ioniser of the quadrupole mass spectrometer are arranged in a vertical manner with the evaporation cells at the bottom of the reaction chamber and the molecular beams directed vertically upward toward the substrate. In this "vertical" evaporation system the position of the beam sources can be properly adjusted for a good overlap of the molecular beams at the substrate surface, and the available cell volume may be exploited with maximum charges, thus reducing the frequency of refilling cycles. If on the other hand the source assembly is positioned approximately in a horizontal geometry with respect to the growth surface, a controlled distribution of solid and liquid source materials into the individual cells, inclined either upwards or downwards with respect to the horizontal plane, is required. A distinct advantage of quasi-horizontal evaporation systems is that contamination of the cell charges by falling flakes, which have been deposited on the shutters and inner walls of the vacuum system in previous growth runs, is minimized. Therefore, the horizontal geometry is preferred in several MBE laboratories.

In addition to the components used directly for the MBE growth process, the growth chamber shown in Fig. 3 contains several components for surface-analytical techniques including reflection high energy electron diffraction (RHEED) at glancing incidence to monitor the surface structure, secondary ion mass spectroscopy (SIMS) and Auger electron spectroscopy (AES) to determine the chemical composition of the substrate and film surface. The associated mass spectrometer of the SIMS equipment is primarily employed for monitoring the background gases in the system in addition to the constituent molecular beam species of interest. Movable screens are used to protect the analytical components inside the growth chamber from deposits from the molecular beams. Thus the basic UHV system schematically shown in Fig. 3 can be used for growing high-quality thin films by MBE as well as for studying detailed surface physics phenomena.

The application of some surface-diagnostic techniques with particular reference to the improvement of control during MBE growth of GaAs and $Al_xGa_{1-x}As$ is described in Sect. 2.2.5.

2.2.2 MBE Effusion Cells

Crucial requirements for the design of MBE effusion cells are a low outgassing rate from the cell material and a negligible reaction between the source charge and the cell material, in order to minimize the concentration of impurities in the molecular beams. For cell materials both high-purity graphite and pyrolytic boron nitride (PBN) have been used. The advantages of graphite are its low cost and its ready machinability to a design which allows good thermal stability and accurate temperature control. Unfortunately, graphite is more difficult to outgas and chemically more reactive than PBN (for example a graphite cell cannot be used for the evaporation of aluminium). Therefore, PBN will become the material of choice for special designs despite its high cost and long delivery time. Cells made of PBN cannot be finished to a final shape but must be used in an as-deposited form often resulting in less accurate temperature measurements.

For practical film growth several effusion cells instead of just one have to be used because of the different vapor pressure between, for instance, gallium, aluminium, arsenic, and the required doping elements. Ideally, the molecular beam sources should be true Knudsen cells which contain condensed phase and vapor at equilibrium[14, 15]. For this case, the beam flux at the substrate can be accurately calculated using Eq. (1)[10]. Approximately true Knudsen effusion cells have only been used for surface kinetic studies. However, the formation of a classical molecular beam is more sophisticated than necessary in an MBE experiment, and the beam intensity from a small aperture is extremely low. Therefore, effusion cells approximately fulfilling Knudsen conditions have been used only for surface kinetic studies of III–V compounds[9, 19, 27], effusion studies[28], and as dopant cells. The orifices of effusion cells containing doping elements can be sufficiently small to ensure that molecular flow is maintained also during film growth. Ilegems et al.[29, 30] and Wood and Joyce[22] have shown for this geometry that the measured impurity concentration versus cell temperature agrees well with a theoretical curve calculated from the known film growth rate, orifice geometry, and source to substrate distance (unity sticking coefficient of the dopant assumed). Recently, Shen[31] has compared the calculation for the beam intensity distribution of an ideal Knudsen-type effusion cell with experimental data and has found excellent agreement. In addition, he measured the intensity distributions from an effusion cell with a single large opening. The problems of volatile species arising from the surrounding of the orifices of multiple high-temperature Knudsen effusion cells have recently been discussed by Chatillon et al.[32].

Non-equilibrium effusion cells of reasonable dimensions with a single large aperture are required for the main constituent elements during practical film growth, in order to obtain large epitaxial areas with uniform thickness and impurity concentration at moderate growth rates. This geometry cannot be calculated, but it actually provides a useful method of collimating molecular beams. Therefore, the emergent molecules must be monitored intermittently using a mass spectrometer, and some kind of feedback system must be employed to regulate the cell temperatures.

Designs of effusion cells for molecular beam epitaxy providing sufficiently high and fairly stable beam fluxes have been described by Foxon et al.[27], Foxon[33], Cho and Arthur[1], and Smith and Pickardt[34]. Most of the commonly used cells are resistance heated. Figure 4 shows schematically a typical MBE effusion cell being free of contamination, which fulfils the requirements of rapid thermal response, uniform heating, and low radiant power loss. The helical tantalum heater wire is supported

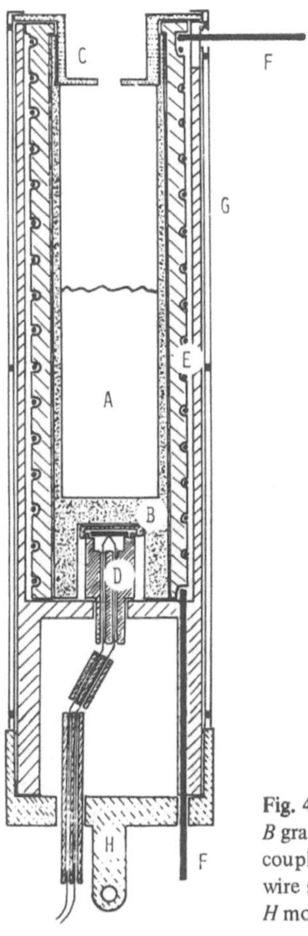

Fig. 4. Schematic design of a typical MBE effusion cell; *A* evaporant, *B* graphite tubular crucible, *C* collimating orifice, *D* alumina thermocouple sheath, *E* concentric alumina cylinders for tantalum heater wire support, *F* tantalum lead, *G* crimped tantalum foil heat shield, *H* molybdenum support for cell adjustment

by two concentric Al_2O_3 cylinders. The specially shaped inner alumina cylinder is lined with reactor grade (or spectrographically pure) graphite which forms an H shaped tubular crucible. In its upper part this crucible contains the charge, whereas its lower part provides a thermally isolated black body enclosing a butt-welded W-Re thermocouple, which passes through low mass alumina tubes and which may touch the bottom of the crucible. With this configuration, the temperature reading at the thermocouple location differs by only ±3 °C from the actual temperature of the evaporation source. The outer alumina cylinder is surrounded by several turns of thin, crimped tantalum foil for an effective thermal isolation of individual cells. Each effusion cell is mounted on an all-molybdenum support with provisions for accurate cell adjustment and rapid dismounting. Tantalum and molybdenum are used exclusively for all support and electrical connections to the cells. Small capacity crucibles with a volume of 2.5 cm^3 for group III and dopant elements as well as large capacity crucibles with a volume of 20 cm^3 for group V elements have been built with the configuration shown in Fig. 4. The diameter of the orifices was chosen to be 3 mm for the small cells and 5 mm for the large ones.

Temperatures, at which these effusion cells operate range from 250 °C to 1400 °C. Precise temperature stability and reproducibility of the individual cells are essential, because temperature fluctuations of ±1 °C can result in molecular beam flux variations ranging from ±2 to ±4%. Since the film growth rate is proportional to the arrival rate of the group III element (for details see Sect. 2.3.1), these temperature fluctuations cause a fluctuation in growth rate by the same order of magnitude. Moreover, in the case of MBE of ternary compounds such as $Al_xGa_{1-x}As$ and $Ga_xIn_{1-x}As$, these fluctuations can produce substantial variations in alloy compositions, for example by about ±4% if the beam intensity of both group III elements fluctuate by ±2% each. For IV–VI compounds, Holloway et al.[35] have calculated beam intensity fluctuations resulting from variations of the effusion cell temperature in a more detailed manner.

If PBN is used as crucible material, the tantalum heater wire can be wound directly on the insulating crucible walls. From the previous discussion it is obvious, however, that there are several difficulties involved in establishing accurate temperature control and measurement since the thermocouple can only be inserted in a rather small recess at the bottom of the as-deposited PBN crucible.

In the author's laboratory the stability of cell temperatures is monitored using programmable temperature controllers with IEC BUS interface connected to a minicomputer. The required temperature and all temperature variations during growth are set by the computer via a digital-analog converter, controlled by a PID controller, and instantaneously read out by a digital thermometer[21]. Temperatures can be set and maintained by the computer in the range from ambient to 1800 °C with an accuracy better than ±0.02%. In order to transmit this accuracy, however, an extremely careful thermal isolation of the effusion cell assembly is a stringent requirement. The cell temperatures can be varied over wide ranges to give growth rates of 0.1 to 10 Å/s for GaAs, and over wide alloy compositions for $Al_xGa_{1-x}As$. The impurity concentration is controlled by adjusting the temperature of the appropriate dopant effusion cell. As a result, MBE can produce nearly any predesigned compositional and/or doping profile perpendicular to the substrate surface by varying the corresponding deposition rate during layer growth. Using this mini-computer device, temperatures and shutter operations can be pre-programmed so that during film growth little interaction of the operator is required.

MBE growth of high-quality semiconductor materials requires a good thermal isolation of the assembly of all of the effusion cells from the rest of the vacuum chamber to minimize outgassing from the walls. This demand is achieved by a liquid-nitrogen-cooled shroud, which surrounds either groups of cells combined on a larger source flange (cf. Fig. 3) or a number of individual cells separately mounted on small diameter flanges. In a group assembly the individual cells can be placed close together and the incident angles of all of the beams impinging on the substrate can be less than 10°. This arrangement makes the distribution of the individual beam intensities very uniform over a large substrate area even at a distance of less than 70 mm between the substrate and the cell orifices. Therefore this set-up is preferred in the laboratory of the author. For the growth of doped GaAs and $Al_xGa_{1-x}As$ the center shroud (Fig. 3) contains gallium, aluminium, and arsenic effusion cells, and the dopant cells are located in the two liquid-nitrogen-cooled shrouds on each side of the main shroud. For the growth of p/n multilayers in GaAs, up to five effusion cells containing the source materials are assembled in the center shroud. This configuration requires a careful heat shielding within the liquid-nitrogen-cooled shroud by tantalum foil in order to prevent temperature changes due to radiation from adjacent cells.

If each effusion cell is mounted on its own small vacuum flange, which contains electrical power and thermocouple feedthroughs, the cells can be completely surrounded with a liquid-nitrogen panel, which isolates the individual cells so that there is definitely no heat transfer between sources. This arrangement, however, requires a larger space between the individual cells resulting in a larger angle of incidence ($>15°$) of the beams impinging on the substrate. In order to make the effusion pattern uniformly directed towards the substrate area, the substrate must be located far ($\geqslant 100$ mm) from the effusion cells. From Eq. (1) follows, however, that large distances between substrate and cell orifices reduce the growth rate. Higher cell temperatures are required to compensate for this effect.

Layer thickness and III–III–V ternary alloy composition may vary over the area of the substrate, if the local angles of incidence of the individual molecular beams are large. Nagata et al.[36)] have used this effect to grow self-aligned three-dimensional $GaAs/Al_xGa_{1-x}As$ multilayers on corrugated structures with 8 μm periods.

Neave and Joyce[37)], on the other hand, designed concentrically mounted Ga and As_4 effusion cells, in order to ensure maximum uniformity of fluxes arriving at the substrate. With this configuration the inner Ga cell orifice is normal to the subtrate. The Ga cell is surrounded by an As_4 cell with an annular orifice.

In most MBE systems each effusion cell is provided with an individual mechanical shutter which allows abrupt changes in beam intensity during continuous growth. The shutters are externally operated either by a direct rotary or linear motion feedthrough or electropneumatically, which can easily be connected to a minicomputer. To minimize the chemical reaction between the source charge and the shutter material, the shutter must be manufactured from a refractory metal.

For many MBE applications growth rates of GaAs and $Al_xGa_{1-x}As$ as well as Al concentrations have been established by trial procedures, using the quadrupole mass

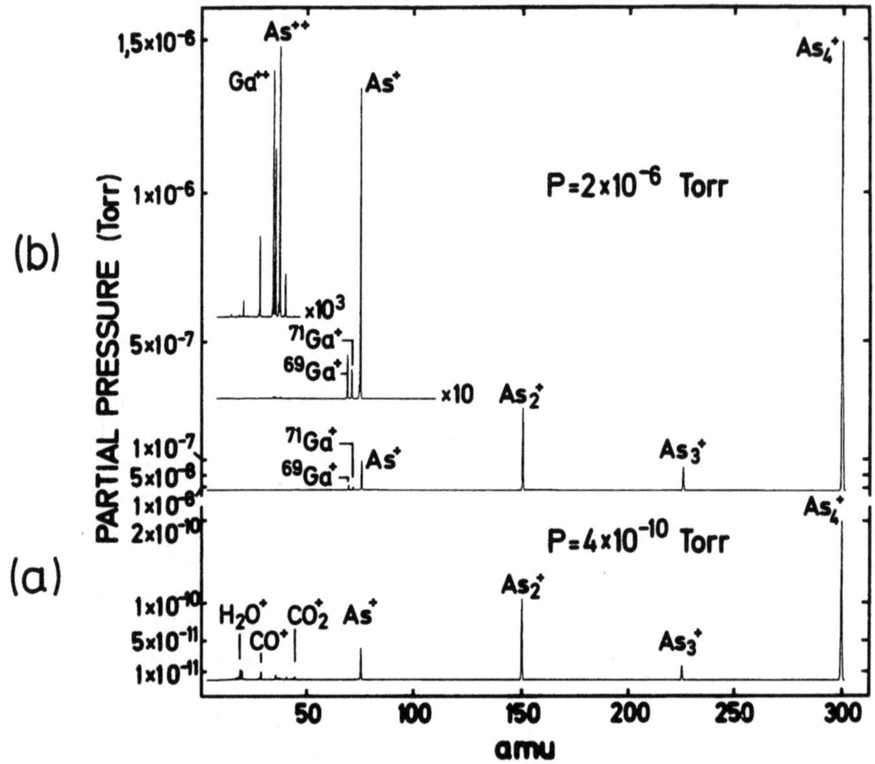

Fig. 5 a–b. Mass spectra of gaseous species in the MBE system shown in Fig. 3 taken in the conventional dc mode; (**a**) prior to epitaxial growth with the *Ga, Al, Be,* and *Si* effusion cells kept at 250 °C and the As_4 cell at 150 °C (before cooling the cryopanels); (**b**) constituent molecular beams and residuals during growth at different amplifications

spectrometer as an aid in setting up the initial conditions. If the molecules emerging from the effusion cells are periodically monitored by a mass filter, the ionisation chamber must be placed with line of sight from the molecular beam sources continuously during a growth run. For the case of GaAs deposition, Fig. 5 shows two typical mass spectra of residual gases for masses up to 300 amu, beyond which no additional peaks are observable at the sensitivity shown. The overall transmission efficiency of the quadrupole mass filter used was uniform within the mass range 1 to 300, so that the spectrum displays directly the relative abundance of the various species. The spectra were taken in the conventional dc mode. Therefore, both background molecules in the vacuum system and molecules from the molecular beam sources contributed to the ion current measured by the mass spectrometer. In order to separate the ac molecular beam signal from the dc signal in the mass spectrometer, it would be more favourable to use a modulated molecular beam technique[38]. This technique, however, is rather complicated *during* epitaxial growth.

The quadrupole mass spectrometer is generally expected to be a promising element-specific detector for molecular beam species. An experimental arrangement proposed by Foxon et al.[38] and schematically shown in Fig. 16 has to be used for quantitative measurements. The ioniser is mounted in the cross-beam mode, having a cryopanel surrounding it, so that molecules passing through without being ionized are condensed and not scattered back. There are, however, several problems encountered in using a quadrupole mass filter to measure molecular beam fluxes quantitatively *during* MBE growth. Among them are the energy-dependence of the probability of producing a detected ion from a deliberate number of neutral species in the ioniser and the ion-extraction process and fragmentation reaction, which are both affected by adjusting the operating parameters of the mass spectrometer. In addition, uncertain long-term stability and instrumental error due to increasing contamination of the ioniser with the evaporants effect this instrument to act not as an *ideal* density-sensitive detector. An application of the mass filter as element-specific sensor for *qualitative* measurements of molecular beam species in closed-loop computer controlled MBE systems has been illustrated by Chang et al.[5], Esaki[39], and by Etienne et al.[40].

After an overnight bakeout, the base pressure of the growth system with all the components installed as shown in Fig. 3 reaches the mid or low 10^{-10} Torr region. As indicated in Fig. 5a, the concentration of the main background impurities H_2O, CO, and CO_2, which can be detected in the reaction chamber in this pressure range before deposition and before cooling the cryopanels, is fairly low. With extended liquid nitrogen cooling, both water and the carbon containing impurities drop drastically. During growth the pressure in the chamber ranges from 5×10^{-7} to 1×10^{-6} Torr. The mass spectrum shown in Fig. 5b indicates that the background pressure is almost entirely associated with arsenic species. This is due to the differences in sticking coefficients between arsenic and the group-III elements (for details see Sect. 2.3.1), which requires an excess arsenic flux during practical growth of III–V compounds.

A versatile method to determine the beam fluxes at the substrate with sufficient accuracy has been described by Cho and Arthur[1]. They calibrated the fluxes of the group-III elements from the film thickness, since the growth rate is directly proportional to the arrival rates of the group-III atoms. The beam fluxes of the dopant elements have been estimated from Hall measurements or Schottky-barrier current-voltage measurements on a series of single layers grown on appropriate substrates. The arsenic flux, which is less critical, was adjusted within suitable limits from the ion-gauge reading and the observation of the RHEED patterns during growth.

For the MBE growth of devices intended for microwave operation and of periodic structures consisting of alternate thin layers of crystalline materials of different composition or doping, an accurate thickness determination on single films is a stringent requirement. This can be achieved only by "post-growth" procedures involving either scanning electron microscopy of cleaved samples or by interferometry using a shadow of the molecular beams cast by a small clamp over the substrate. Since the Ga and Al molecular beams are not coaxial during growth of $GaAs/Al_xGa_{1-x}As$ multilayer structures, separate shadows of the clamp are formed for GaAs and AlAs. This technique enabled Gossard et al.[6] to determine GaAs, AlAs and alloy thickness of the deposited $(GaAs)_n$-$(AlAs)_m$ superlattices separately. Finally, Petroff et al.[41] have carried out a more precise measurement of film thicknesses of these superlattices by transmission electron microscopy. Using a GaP monitor slice as a blank mounted simultaneously on the substrate heater, Wood[42] has determined single film thicknesses by optical absorption at 7800 Å through the GaAs/GaP heterostructure.

Foxon and Joyce[19] and Wood and Joyce[22] have measured the beam fluxes with an inexpensive movable ion gauge, which is first moved into the substrate position to measure the direct beam density and then out of the beam to measure the background density. Although in general only used to ensure reproducibility, a calibration of the ion-gauge can be made absolute for the group-III elements of interest. For a unit sticking coefficient the ion current is directly related to the mass of each element deposited on a thin metal plate in the beam. The amount of, for example, condensed Ga is determined by the change in weight of the plate. For the group-V elements arsenic and phosphorus again, however, only a relative calibration is possible because of their non-unity sticking coefficients.

The first optical method investigated for monitoring the flux emitted from effusion cells during MBE was atomic absorption spectrometry ("AAS"). Kometani and Wiegmann[43] discussed the feasibility and limitations of AAS as a non interfering, quantitative method for monitoring both Ga and Al molecular beam fluxes during deposition of $Al_xGa_{1-x}As$. Even after several fundamental improvements, however, only changes as large as 5 °C in the Al effusion cell temperature could be readily detected by AAS while growing $Al_{0.2}Ga_{0.8}As$[43].

A very promising optical method for monitoring molecular beams is based on the principle of atomic beam emission spectroscopy. The detection unit of the commercially available Sentinel 200[44] is based on the Electron Impact Emission Spectroscopy (EIES) technique. The sensor of this instrument is placed immediately adjacent to the substrate heater block. An electron beam with a rectangular cross section (175 V, 6 mA) interacts with evaporant atoms that enter the probe. The atoms are excited by collision to unstable states resulting in the emission of radiation which is characteristic of the particular atomic or molecular species. After the emitted light passes through a light pipe and appropriate narrow band pass filter or a monochromator, the intensities of the emitted lines are measured with a photomultiplier. For Ga a line at 3184 Å and for Al a line at 3093 Å are usually selected for monitoring molecular beams. If the electron beam current and energy are held constant, the photon flux is directly proportional to the evaporant number density, which is – in the case of group-III elements – related to the deposition rate.

2.2.3 Source Materials

For source materials to generate the molecular beams, the pure group-III and -V elements and additional separate dopant elements (Be, Si, Ge, and Sn, respectively) are generally used. All starting materials should be 99.9999 + % pure. Only in the case of Be has it turned out to be difficult to obtain material commercially having a nominal impurity concentration $\leqslant 1$ ppm. The thermal molecular beams of all group-III and dopant elements consist of monoatomic species, while the beams emerging from sources of elemental group V materials are always composed of tetrameric molecules (P_4, As_4, Sb_4). Since the evaporation coefficients of elemental phosphorous and arsenic differ from unity[45], the beam fluxes can be estimated accurately only by direct measurements using a calibrated mass spectrometer. Several authors have reported[26, 46] that the amount of P_4 consumed during the MBE experiment is apparently much smaller ($\sim 1/10^3$) than the evaporated quantity as calculated from the vapor pressure of phosphorus under Knudsen conditions.

Additional useful sources of group-V-element molecular beams are the III–V compounds (GaAs, GaP, InP) themselves. They provide constant, well-determined beam fluxes as long as a minimum amount of the group-V element, as determined by the phase diagram, is present in the group-III-rich solution. When solid GaAs is heated to above $\sim$635 °C, which is the limiting temperature of congruent evaporation[38, 47], the volatile arsenic evaporates preferentially leaving a Ga-rich solution behind within the effusion cell. Thus, for the evaporation of GaAs, only the Ga-rich region of the binary Ga-As liquidus curve as given in Fig. 6 is of interest, and the vapour pressures of Ga, As_2, and As_4 can easily be determined experimentally[38, 47].

When a III–V compound is used as source of the group-V element, the emerging beams are principally composed of dimeric molecules. Foxon et al.[48] have found that the As_2/As_4 ratio starting from GaAs is $\geqslant 10^4/1$ as is the P_2/P_4 ratio emanating from GaP. From InP, however, the P_2/P_4 ratio is only approximately 10/1.

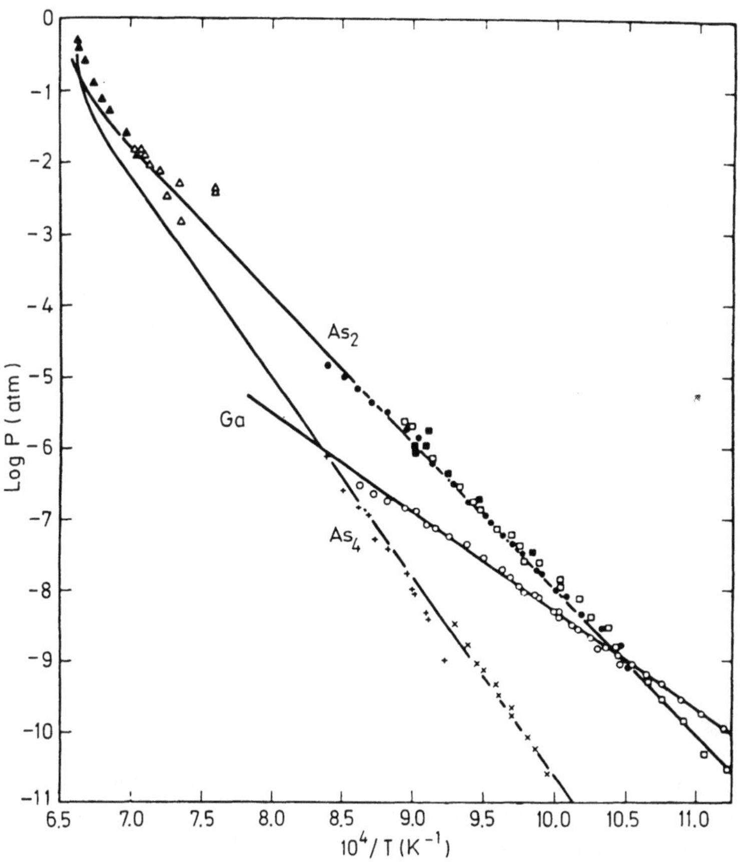

Fig. 6. Equilibrium pressures of As_2, As_4, and Ga over GaAs along the Ga rich portion of the binary Ga-As liquidus compared with calculated curves (from Ref. 47)

The advantages and disadvantages associated with the application of either elemental arsenic or polycrystalline GaAs as arsenic source have been carefully considered by Cho and Arthur[1]. The advantage of using the GaAs charge over the pure arsenic charge is that oxidation of the compound source is much less extensive than with the pure arsenic during handling and exposure to air. Removal of the various oxidation products of arsenic requires a heating of the source above its operating temperature for some period of time prior to growth. Despite these difficulties, the use of elemental arsenic sources produces purer films with a lower doping background, because in general the commercially available undoped polycrystalline GaAs contains Si in a concentration of about 10^{16} cm^{-3}. Cho and Reinhart[49] have evaluated that the use of those GaAs ingots produces unintentionally doped n-type GaAs film with a background donor concentration N_D-N_A ranging from 5 to 8 x 10^{15} cm^{-3}. When pure gallium and pure arsenic were used as sources under otherwise similar growth conditions, however, the background carrier concentration was drastically reduced now yielding p-type material with acceptor concentrations in the 10^{14} range.

When the MBE technique is extended to grow GaAs planar electrically isolated areas on the substrate[50] or to achieve a "writing" of patterned GaAs epilayer structures using mechanical masks[51–53], a nearly concentric beam of the constituent elements impinging on the substrate is required. This geometry is easily achieved by using a single GaAs containing source, which can provide a sufficient amount of both Ga and As_2 species in a limited temperature range. Figure 6 shows that Ga partial pressures of 2 x 10^{-4} to 7 x 10^{-4} Torr within the cell are obtained at 830–910 °C along with As_2 pressures of 10^{-2} to 10^{-3} Torr. For a source to substrate distance of 65 mm this yields Ga fluxes of about 2 to 8 x 10^{14} atoms/cm^2 · s at the substrate or approximately 0.2 to 0.9 monolayer/s corresponding to a limited growth rate of around 0.5 μm/h.

For the geometry used by the author with separate effusion cells containing Ga, Al, and elemental arsenic the required fluxes of the species for growth rates between 0.3 and 1 μm/h are obtained at the following temperatures:

10^{14}–10^{15} Ga/cm^2 · s at 930–980 °C;

10^{14}–10^{15} Al/cm^2 · s at 1070–1130 °C;

10^{15}–10^{16} As_4/cm^2 · s at 300–330 °C.

2.2.4 Substrate Processing

The preparation of the growth face of the substrate from the polishing stage to the in-situ cleaning stage is very important for molecular beam epitaxy as it is for any other epitaxial growth technique. This is so because the effects of the substrate-active-layer interface on the device performance are significant for rather thin films ($\leqslant$1 μm without buffer layer), the growth of which is the real strength of the MBE technique. In order to obtain uniform and smooth MBE layers of good electrical and optical quality, in situ cleaning and in situ assessment of the initial substrate surface within the growth chamber are essential.

Except for some early studies[54–56] most of the MBE growth of GaAs and $Al_xGa_{1-x}As$ has been performed on (100) oriented (±0.1°) GaAs slices being 200 to 400 μm thick. In many cases it is very useful to deposit the epitaxial films simultaneously on several substrates; a particularly doped GaAs wafer for the desired device together with a semi-insulating (mostly Cr-doped) GaAs and an n^+-GaAs slice as monitor for thickness, Hall measurements, and Schottky barrier current-voltage characteristics, respectively.

The general procedure for preparation of (100) GaAs substrates involves polishing with diamond or natural corundum paste to remove the saw-cut damage and etch polishing on an abrasive-free lens paper soaked with a sodium hypochlorite or bromine-methanol solution[57] leaving a mirrorlike finish. The substrate is then boiled in hydrochloric acid, rinsed in water, and free etched in a $H_2SO_4/H_2O_2/H_2O$ solution (3 : 1 : 1). Finally, the wafer is flooded with double-destilled water to stop the etching, and the residual liquid is blown off the surface with a jet of filtered nitrogen just before mounting on the Mo heating block in the UHV system. At this stage, the GaAs substrate will be contaminated by principally carbon and oxygen impurities, their relative surface concentrations depending on the precise etching procedure used (Fig. 7). Cho and Arthur[1] have found that the amount of oxygen can be increased and the amount of carbon reduced by washing the etching solution away with water before the surface is exposed to laboratory air. Using this technique the GaAs surface is passivated and thus fairly stable to further carbon contamination during loading into the MBE system. This is an important result because the oxide can be easily removed by heating in vacuum to 525–535 °C[49; 58, 59], a temperature well within the range of congruent evaporation.

Immediately after etching, the substrate is soldered to a Mo plate with liquid In under dust-free conditions, in a simple modification of the method described by Cho[60] and Cho and Arthur[1]. They have found that the surface tension of liquid In (or Ga) is sufficient to hold the substrate in either a horizontal or vertical position even when heated to above 600 °C. The thin layer of molten metal provides good thermal contact of the substrate to the Mo plate, and a uniform heating over the entire substrate area can be achieved. If care is taken during mounting, any stress on the substrate producing dislocations in the epitaxial layer along the [110] directions can be eliminated. Wood et al.[61] have found that a dissolution of the GaAs in the In solder at the growth temperature and the reprecipitation of $In_xGa_{1-x}As$ on the back surface of the substrate can be avoided by a thin (1000–2000 Å) SiO_2 film sputtered on the reverse of the substrate before processing.

In order to change the substrate after each growth run, most of the MBE systems described in the early literature required that the growth chamber be brought up to atmospheric pressure using dry nitrogen gas. During this time-consuming procedure, moisture and contaminants from the laboratory atmosphere can readily be adsorbed onto the inner walls and on the effusion cells. This problem of contamination can be overcome and a substantial reduction of time loss in changing samples can be achieved by a particular sample exchange load-lock system as used for the MBE system shown in Fig. 3. Via this interlock system the Mo mounting plate with the freshly prepared substrate is remotely exchanged with a plate holding the processed substrate from the previous growth run. After removing the old substrate plate the new plate is

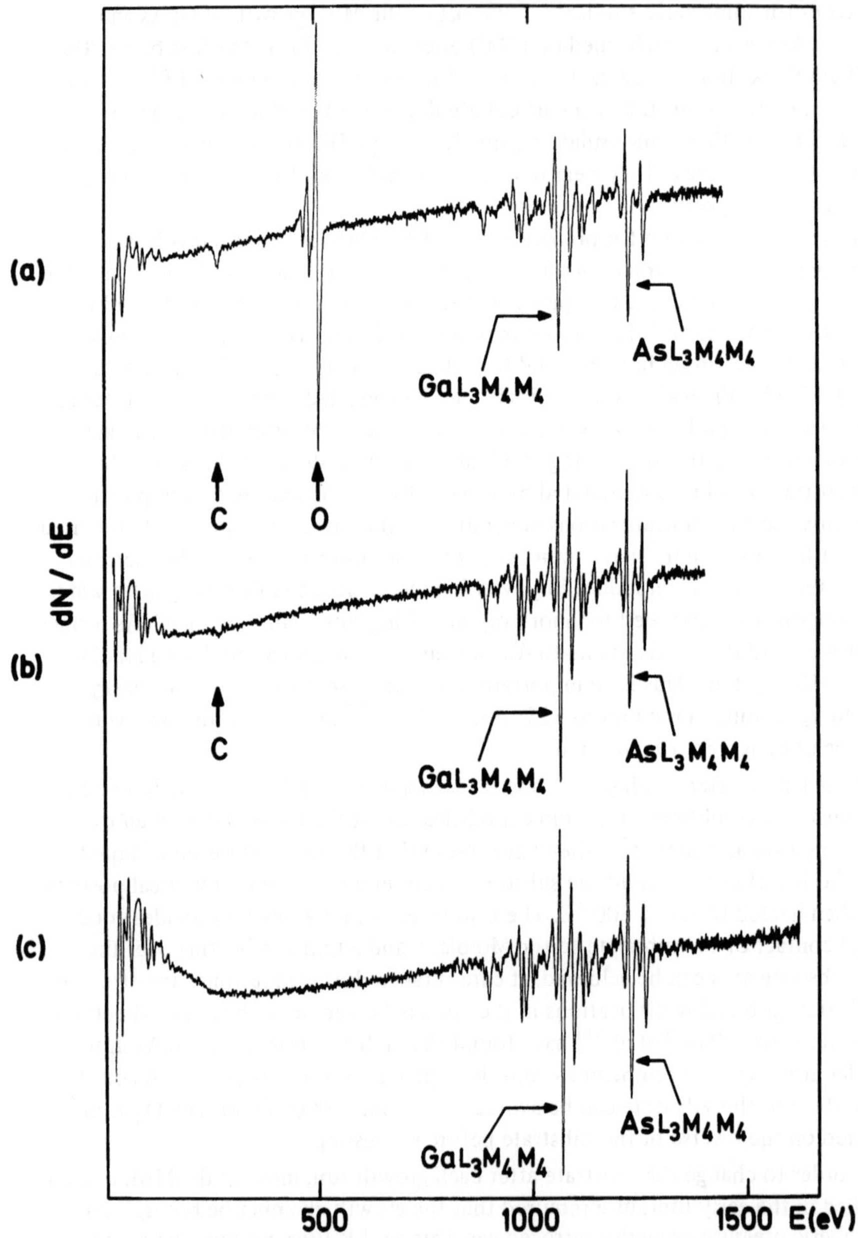

Fig. 7 a–c. Auger spectra of variously treated (100) GaAs substrate surfaces prior to MBE growth taken (a) after etching in $H_2SO_4/H_2O_2/H_2O$; (b) after heating of the chemically etched wafer to 540 °C; (c) after additional Ar^+ sputter-cleaning of the chemically etched surface and annealing at 350 °C

attached to the transport probe by inserting it through a rapid access flange on the introduction chamber. When this chamber has been evacuated to $p \leqslant 10^{-7}$ Torr, the substrate and transport probe are driven into the growth chamber by a special magnetic linear-rotary motion drive. The plate is then fixed to the internally heated Mo block which remains always inside the growth chamber. Finally, the transport probe is removed from the growth chamber, and the all-metal isolation valve between main vacuum chamber and introduction chamber is closed. During sample exchange the cryopanels in the growth chamber may be held at liquid-nitrogen temperature, and the effusion cells may be kept at temperatures not far below their usual operating temperatures. Five minutes after completing the sample change the pressure in the growth system is in the low 10^{-9} Torr range again, and a new growth run can be started rapidly without any change in the growth conditions.

With reference to Fig. 3, the internally heated Mo block is attached to a standard off-axis rotary manipulator mounted on a 100 mm I.D. flange which also contains power and thermocouple feedthroughs for heating of the substrate during growth. The distance from the manipulator axis, which is also the center of the growth chamber, to the substrate is 65 mm. By rotation of the manipulator the sample can be brought to the proper position for sample exchange, to face the effusion cells, which is the position for growth at a cell-substrate distance of 70 mm and for reflection electron diffraction at glancing incidence, or to face the particular components for sputter cleaning and/or surface composition analysis.

The temperature of the substrate mounted on the remotely exchanged Mo plate is measured with a butt welded W-Re thermocouple passing through the Mo heater block and being rigidly fixed 1 mm below the surface to which substrates are attached. Using this arrangement it is not possible to measure the absolute temperature of the substrate accurately. However, it is possible to obtain and maintain a given temperature in a reproducible manner, even when using the sample exchange device, and this is all that is required for successful film growth. In practice, the temperature measured within the heating block and the true temperature of the substrate surface can be calibrated in the range of interest using the well-established temperature of 525–535 °C[51, 58, 59] at which the passivating oxide film on GaAs evaporates, and the well-known temperature of 635 °C[38, 37] indicating the limit of congruent evaporation of GaAs. A more sophisticated method of calibrating the temperature of substrates attached to a heater block has been described by Matysik[62] and by Delhomme et al.[63]. Neave and Joyce[64] have recently developed an internally heated Mo block providing uniform heating over large areas at temperatures ranging from 80 °C to 600 °C with the measured values being correct to ±5 °C. This device is particularly useful for surface kinetic studies with special reference to substrate temperature effects.

In addition to the method of conductive heating described above, several other facilities of controlled heating of substrates during MBE experiments have been described in the literature including radiant heating[42] and direct resistance heating[38]. For radiant heating, the heater consists of a vitreous silica plate coated on the back surface with a sputtered Ta film. The substrates are attached to the front of the plate using light clamps and the conductive Ta film is resistively heated. Direct resistance heating can be accomplished by passing a current through rather thick rectangular shaped substrate bars. This method permits accurate control and measurement of substrate temperature by using a thermocouple placed in a hole drilled right through the narrow face of the bar. Thus, it is particularly useful for research applications, when only the substrate itself is allowed to be at elevated temperature.

The in situ characterization of the substrate surface prior to epitaxial growth is carried out using Auger electron spectroscopy (AES) and reflection high energy electron diffraction (RHEED). Figure 7 shows Auger spectra of variously pretreated (100) GaAs substrate surfaces obtained with a cylindrical mirror analyser (CMA). Spectrum (a) was taken after a typical chemical etching in $H_2SO_4/H_2O_2/H_2O$, rinsing in water, loading into the growth chamber. Besides the complex high-energy structure of Ga and As, a large peak due to surface oxygen at 510 eV and a smaller one due to surface carbon at 272 eV are found. The outermost oxygen layer is intentionally produced by the special etching process used leading to a passivated GaAs surface which is fairly stable against further carbon contamination. As indicated from spectrum (b) in Fig. 7, the oxygen can be removed by simply heating the substrate to ~535 °C in vacuum[51, 58, 59], and so a nearly clean surface with only a few percent of a carbon monolayer is easily produced. Cho and Panish have found[58] that carbon contamination at this low level does not have any noticeable effect on epitaxial growth and on the quality of the resulting films.

In a more thorough investigation using Raman spectroscopy, Abstreiter et al.[65] have recently re-evaluated the effect of carbon contamination on MBE grown GaAs films. With the Raman technique applied in a backscattering geometry on as-grown (100) surfaces, only the longitudinal (LO) phonon is allowed from the selection rules. Typical examples of Raman spectra observed from GaAs films (thickness $\leqslant 1\ \mu m$) grown on variously pretreated substrate surfaces are shown in Fig. 8. A deviation from the Raman scattering selection rules and the appearance of the forbidden TO mode depend strongly on the pretreatment of the substrate surface. The strong mode observed in some MBE films (spectrum d in Fig. 8) grown on well-oriented (100) GaAs substrates clearly demonstrates that the orientations of at least parts of the films are different from the expected (100) direction. The quality of the grown GaAs

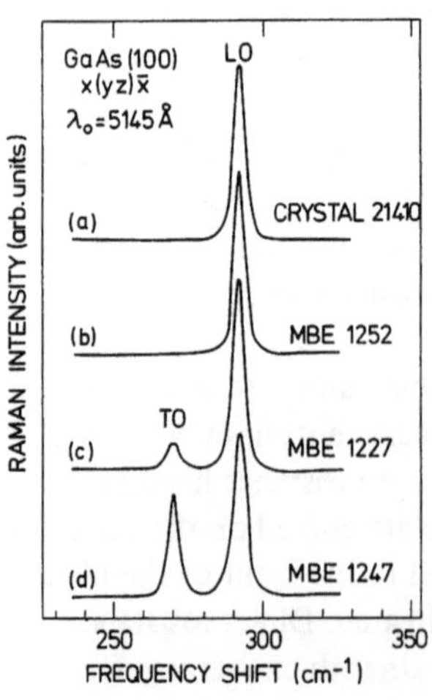

Fig. 8. Raman spectra obtained in backscattering geometry from GaAs films grown on variously pretreated (100) GaAs substrates

crystal is perfect with no TO phonon mode present, only in the case when the residual carbon contamination on the substrate surface is less than 0.1 monolayer, which is often the limit of detection of AES measurements.

Abstreiter et al.[65] have suggested that the strength of the TO phonon line is directly correlated with sub-monolayer quantities of carbon on the initial growth surface. Starting with a contamination $\geqslant$0.2 monolayer carbon, a twinning of the growing GaAs film can be observed even in the RHEED pattern and the sample shows a strong "forbidden" TO phonon, with an intensity about 2/3 of that of the LO phonon (spectrum d in Fig. 8). In addition, carbon contamination in excess of 0.2 monolayer has an important effect on growth topography. Figure 9 shows two scanning electron micrographs of successive steps of GaAs growth on a carbon contaminated substrate surface. The micrograph in Fig. 9a was taken after the deposition of an equivalent of about 200 Å. A three dimensional nucleation has occurred yielding irregular three-dimensional centers and discrete three-dimensional crystallographically shaped growth centers in which the deposited material is concentrated. Even after the deposition of more than 5000 Å (Fig. 9b) the surface of the grown GaAs film has not smoothed out, and distinct depressions with rectangular-shaped edges extend from the surface into the epitaxial material toward the substrate-layer interface. For a carbon contamination in between 0.2 and 0.1 monolayer there is no observable effect in the RHEED experiments and the microscopic appearance of the film surface is smooth, although a small forbidden TO line is still evident in the Raman spectrum c shown in Fig. 8.

Laurence et al.[66] demonstrated recently that even a carbon coverage as low as $\sim 6 \times 10^{-2}$ monolayer induced surface faceting when heating the GaAs substrate in a flux of arsenic. The temperature at which faceting was observed depends markedly on the carbon coverage measured by AES after oxygen desorption. With $\sim 6 \times 10^{-2}$

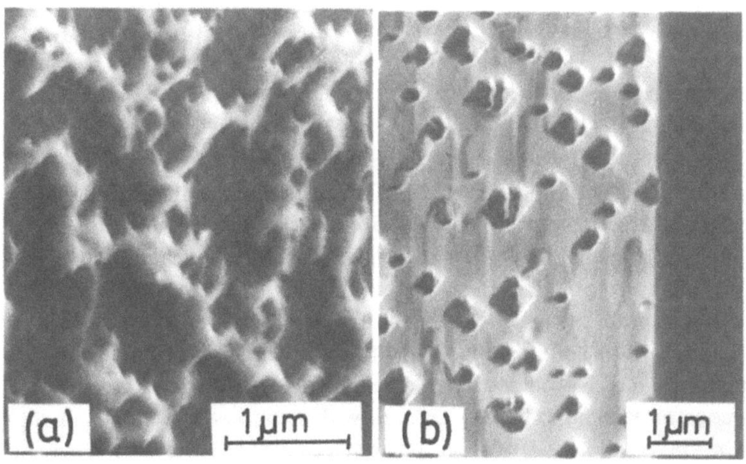

Fig. 9a and b. SEM micrographs of GaAs films grown on carbon contaminated (100) GaAs substrate surfaces; **(a)** three-dimensional island growth after depositing an equivalent of 200 Å GaAs; **(b)** formation of depressions with regular-shaped edges after continuous deposition of 5000 Å GaAs

monolayer carbon, (411) facets appear on As-stabilized (100) GaAs surfaces heated in an As_4 flux to a critical temperature ranging from 575 to 585 °C.

The nature of the misorientation or faceting of the (100) GaAs films grown on carbon contaminated surfaces which is responsible for the appearance of the forbidden TO mode in the Raman spectra is not yet clear. Additional X-ray and electron diffraction studies are necessary to get more detailed insight into the microscopic growth steps. The effect just described strongly suggests, however, that the substrate preparation is the most important step for the successful growth of high-quality thin films by MBE. A carbon contamination on the substrate surface strongly affects Hall mobilities of very thin (>0.5 μm) layers, because only on carbon-free surfaces layer growth can propagate via a two-dimensional step mechanism. The required purity of the substrate surface can be achieved by either an extremely careful chemical etching, which, however, might vary from sample to sample, or in a more reproducible manner by an additional Ar^+ ion sputter cleaning and annealing prior to epitaxial growth. This last cleaning procedure is controlled with AES (Fig. 7) and with RHEED (see Sect. 2.2.5.2) which monitors the recovery of the structure of the sputter-damaged surface after annealing.

Since the surface contamination is strongly dependent on the substrate preparation technique used, and since heat treatment is not effective in carbon removal, it is advisable to have some facility for additional ion-sputter-cleaning in the MBE growth chamber. Consequently, sputtering is routinely done in many laboratories, in order to insure reproducibility from run to run. In the MBE system shown in Fig. 3 a differentially pumped ion gun yielding low energy ions is applied for this purpose. Using this technique, the ambient pressure around the target can be kept in the low 10^{-9} Torr range during operation of the gun, in order that the adsorption rate of residual gas species may be smaller than the erosion rate. The angle of incidence of the ion beam to the substrate surface is 20° and the beam can be scanned to achieve uniform removal of contaminants over large substrate areas. In order to keep the surface damage from the ion beam only a few atomic layers deep, Ar^+ ions of low energy (500–1000 eV) are used at low current densities ($\leqslant 1\ \mu A/cm^2$) for prolonged sputtering times ($\geqslant 2$ h). It is estimated that under these conditions less than 0.1 μm/h of GaAs are removed from the previously chemically etched surface.

Before starting growth, the crystallinity of the sputter-damaged surface must be recovered by annealing the GaAs sample at ~350 °C in a beam of arsenic. During this treatment the RHEED diffraction pattern changes from amorphous background diffraction to faint streaks. Immediately after annealing the substrate is heated to between 560 and 600 °C for up to 20 min in a As_4 flux of 10^{14}–10^{15} molecules/cm$^2 \cdot$ s. This final heat treatment produces an As-stabilized surface which shows a streaked C (2 x 8) or (2 x 4) RHEED pattern (cf. Fig. 13) with very little intensity modulation along the length of the reciprocal lattice rods, indicating that the substrate surface is now strongly reconstructed and extremely smooth with steps down to atomic dimensions. Often the epitaxial layers grown on sputter-cleaned substrates show a somewhat smoother appearance than those grown on chemically etched substrates when examined with a scanning electron or phase contrast microscope. The influence of substrate preparation on Hall mobilities for very thin GaAs layers has been thoroughly investigated by Cho et al.[67]. The authors have illustrated the effect of the substrate-active-layer interface on the performance of GaAs metal-semiconductor field-effect transistors (MESFET's) by growing wafers either on ion-sputter cleaned surfaces without a buffer layer or on chemically etched substrates with a buffer layer.

A second cleaning technique to remove carbon in situ from a chemically etched GaAs surface has been developed by Wood[42)]. This method involves heating of the substrate at 580–610 °C in ambients ($\sim 10^{-6}$ Torr partial pressure) of water vapor or oxygen. Again, any oxide formed during this treatment can easily be removed by heating in vacuum.

The cleaning technique actually chosen depends strongly upon the contaminants involved and on the device structures being grown. Ideally, any in situ cleaning in addition to a careful chemical etching should be suspended. Currently, in the author's laboratory a heat treatment of the substrate in a beam of As_4 ranging from 10^{13} to 10^{15} molecules/cm^2 · s for up to 30 min at 560 to 600 °C is sufficient to produce an As-stabilized surface exhibiting pronounced elongated bulk diffraction spots and a rather weak (2 x 4) or C (2 x 8) reconstruction. Differently doped substrates require different time. This treatment can achieve a surface with residual carbon and oxygen impurities at levels of less than a few percent of a monolayer as monitored by AES.

Indium phosphide is the only other III–V substrate material for which the process of substrate preparation before MBE growth has been described in some detail in the literature by Farrow[11)], McFee et al.[68)], and by Asahi[69)]. Farrow[70)] has found that in the case of InP, simple heat cleaning is ineffective since close to the congruent evaporation temperature the rate of congruent removal is only $\sim 10^{-5}$ monolayers/s. Thus, the most effective and only reproducible way of in-situ cleaning involves low energy Ar^+ ion sputtering followed by annealing for up to 1 h at $T < 400$ °C in a beam of the group V element.

Finally, the primary electron beam used to monitor the surface quality by AES and/or RHEED can itself enhance the recontamination rate for both carbon and oxygen on the surface of GaAs and $Al_xGa_{1-x}As$[1, 20, 49)]. This effect can be reduced if both diagnostic techniques are not extensively used during the background pressure of reactive gases being greater than 5×10^{-10} Torr. For practical film growth, however, a substrate on which the grown film is to be used for device fabrication should not be monitored by either technique involving an electron beam onto the growing surface. One possible way out of this dilemma is to assume that the particular cleaning procedures applied have been successful, another is to perform the surface analysis on a second substrate which has been identically prepared.

2.2.5 Application of Surface Diagnostic Techniques During MBE

The growth technique of molecular beam epitaxy has emerged from surface studies of the interaction of thermal molecular beams with solid surfaces under UHV conditions. In the systems GaAs-Ga-As_2 and GaAs-Ga-As_4 measurements of thermal accommodation coefficients, surface lifetimes, sticking coefficients, desorption energies and reaction orders have been made by Arthur[4, 17, 71)] and by Foxon and Joyce[72–74)] (for details see Sect. 2.3). In order to get reproducible results from the kinetic measurements, the authors had to make a detailed assessment of the conditions of the starting substrate surface by using appropriate surface-analytical techniques. Such techniques included Auger electron spectroscopy (AES), reflection high energy electron diffraction (RHEED) and – more recently – secondary ion mass spectroscopy (SIMS).

From this origin of molecular beam epitaxy at least one or two of these techniques have been continuously applied in situ to ensure that the desired conditions have been reached before and are being maintained during crystal growth[20)]. An incorporation of the surface diagnostic techniques AES, RHEED, and SIMS directly into the growth apparatus has contributed important details of the fundamental processes, which lead to the formation of high-quality monocrystalline GaAs films by MBE. The results obtained by these techniques can thus enable the operator to establish, to control, and to optimize the conditions for successful growth in situ. This point will be demonstrated with the aid of some selected examples.

In addition, the MBE technique enables us to prepare flat, well-ordered, clean semiconductor surfaces of distinct orientations and to alter the surface composition in a well-defined reproducible manner. Both features will be essential for more sophisticated surface experiments in the future, which may include low energy electron diffraction (LEED), low energy electron loss spectroscopy (LEELS), photoelectron spectroscopy (UPS and XPS), and Raman scattering. A brief outline of some possible applications of these experiments has been given in a recent paper[75)].

2.2.5.1 Surface Studies by Auger Electron Spectroscopy (AES)

Auger electron spectroscopy (AES) is a non-destructive analytical method for the determination of the chemical composition of the outermost atomic layers of surfaces with a detection sensitivity ranging from 0.1 to 0.01 monolayer. Details of the AES technique have been extensively discussed, for instance, by Chang[76)]. In the MBE system shown in Fig. 3, the Auger measurements have been obtained with a cylindrical mirror energy analyser (CMA) using a 5 keV coaxial incident electron gun which was regularly operated at 3 keV. The primary electron beam was focussed down to 100 μm diameter. For the normal dN/dE spectra the primary current was reduced to less than 1 μA, since higher currents could cause alterations in the chemical and structural properties of the GaAs surfaces. In order to get more quantitative information on the surface Ga and As concentrations, direct N(E) spectra can be taken with the same CMA equipment by using the brightness modulation technique[77)] and fairly low primary currents in the range of 10^{-8} to 10^{-9} A.

During MBE experiments the Auger technique has been extensively used for several different purposes. Among them are:

(i) Characterization of the initial substrate surface. The considerable value of AES in determining the effectiveness of in situ cleaning of the substrate surface prior to film growth has been demonstrated in Sect. 2.2.4.

(ii) Experimental verification of surface accumulation of dopant elements during epitaxial growth. For the case of doping MBE GaAs with the n-type dopant Sn, AES has provided definite experimental evidence[78)] that the deviation from the ideal doping profile[60)] is caused by an enrichment of Sn in the outermost atomic layers during deposition. This distinct behaviour of Sn will be more fully discussed in Sect. 2.3.2.

(iii) Determination of the relative ratio of the constituent elements of reconstructed surface structures. The AES measurements have confirmed the suggestion of Arthur[71)] that surface-structural changes are correlated with gain or loss of arsenic from the (100) GaAs surface, since As atoms are only weakly bonded on the surface in comparison to Ga atoms. Some of the results obtained on the relationship between surface stoichiometry and reconstruction will be discussed in combination with RHEED studies in Sect. 2.2.5.2.1.

(iv) Determination of the bulk chemical composition of ternary and quaternary III–V films. So far, the use of AES for accurate quantitative analysis has been fairly limited, because the Auger current, I, of an element depends on Auger transition probabilities, on instrumental parameters, and on a number of matrix effects which include variations in the chemical environment, in the mean free path of Auger electrons, and in the fraction of the primary beam being backscattered by the target. The uncertainties in calculating and measuring these matrix effects have led most quantitative AES measurements to rely on comparison with known standards. Even then, however, a good accuracy can be expected only when the standard and the unknown crystal are quite similar. Motivated by the technological importance of GaAs and AlAs, the $Al_xGa_{1-x}As$ system is exceptionally well studied by AES because the principal Auger peaks of all three components Al, Ga, and As do not interfere and because all three elements have both low energy (30–65 eV) and high energy (1000–1400 eV) Auger transitions which have different escape depths ranging from 5 to 17 Å[76)]. Their relative intensities can thus provide some indication of concentration gradients at the surface. Hammer et al.[79)] and Arthur and Le Pore[80)] have performed AES analyses of the Al-Ga-As system based on calibration against internal standards. For $Al_xGa_{1-x}As$ samples with $0 \leqslant x \leqslant 0.95$ Arthur and Le Pore obtained a long term composition reproducibility of ±3% or better. These measurements have been correlated with near-band-gap absorption and luminescence studies[81)].

(v) Analysis of composition as a function of depth into the grown crystal. Ludeke et al.[82)] have used the profiling technique of AES coupled with argon ion sputtering[83)] to examine MBE grown $GaAs/Al_{0.25}Ga_{0.75}As$ superlattice structures; Chang and Koma[84)] have measured the coefficients of interdiffusion between GaAs and AlAs in similar structures. While Ludeke et al. could observe the expected periodic variation of the Al (and Ga, resp.) Auger peak in a 16 1/2 period superlattice with single layer thicknesses as small as 50 Å, the intensity of modulation decreased with depth into the crystal. This behaviour has been attributed to the microtopography induced by ion bombardment at the bottom of the elliptically shaped crater created on the target[82)]. A second inherent limitation in high-resolution depth analysis of layered structures is imposed by the escape depth of the Auger electrons which is probably 10–17 Å for the range of energy chosen by the authors.

(vi) Taking low-energy electron loss spectra (LEELS or simply ELS) from the as-grown MBE material by recording the second derivative of the energy spectrum of electrons inelastically backscattered into the CMA analyzer. This kind of LEELS technique has been developed mainly by Ludeke et al.[85–87)] using primary electrons of an energy ⩽200 eV. Loss spectra of a semiconductor surface involving valence electrons yield details about interband transitions between bulk states and between surface states. Furthermore, the loss spectrum due to excitations of core electrons into empty states can be obtained. From the combined information contained in those spectra, the authors were able to deduce details of the energy-level schemes of surface states for III–V compounds and some of their oxides and to compare them with appropriate theoretical calculations and with results obtained by photoelectron spectroscopy.

2.2.5.2 Surface Studies by Reflection High Energy Electron Diffraction (RHEED)

Reflection High Energy Electron Diffraction (RHEED) is probably the most important and most effective control technique for MBE growth. The RHEED technique provides a very sensitive yet simple diagnostic tool to observe changes in the structure of the outermost layers of atoms in a crystal *during* growth as a function of growth parameters, when used at glancing incidence. It is possible to obtain information on surface structure, microstructure, and smoothness from RHEED[88]. The method has played an important role in the development of molecular beam epitaxy[89].

In the MBE system shown in Fig. 3, a well-collimated monoenergetic electron beam is incident at a grazing angle of about 1° on the substrate surface orthogonal to the molecular beam paths. The accelerating voltage of the primary electron beam is typically in the range 10–40 keV. However, the electron beam may also be obtained from a medium energy electron gun (1–3 keV), thus providing quite an inexpensive diffraction system more properly called "RMEED". With these energies there might arise some problems with beam collimation, which, however, do not affect routine evaluation during MBE growth. In either case the penetration depth of the incident electron beam is restricted to the first few atomic layers of a crystal, since the decisive energy component normal to the surface is in the order of 30–150 eV even though the total energy of the beam is 1–50 keV. A smooth crystal surface thus acts as a two-dimensional grating and diffracts electrons as discrete beams, which form a characteristic pattern on a fluorescent screen placed perpendicular to the diffracting surface and incident beam azimuth. This geometry allows the molecular beams to be on and the crystal continuously growing while the diffraction pattern is observed. During epitaxial growth on (100) oriented III–V substrates the primary electron beam impinges on the growing film in different, approximately [110] directions. The symmetry and spacing of the reciprocal lattice may be determined very easily from the diffraction pattern. In order to obtain the whole translational symmetry of the surface, two or more different azimuthal patterns must be taken. In addition, RHEED provides information on surface topography which is of considerable value in studies of the early stages of epitaxial growth.

The type of RHEED pattern obtained from a smooth single crystal surface is a series of streaks which are oriented in a direction perpendicular to the surface being observed. As yet, however, a satisfactory explanation as to the origin of streaks in a RHEED pattern has not been given. When examining published work, it is found that streaked patterns are generally believed to be indicative of flat, well-defined single crystal surfaces, the Ewald construction of which is shown in Fig. 10. This idea is supported by two successive RHEED patterns before and during epitaxial growth of GaAs which are shown in Fig. 12. The spotty pattern in Fig. 12a, obtained at room

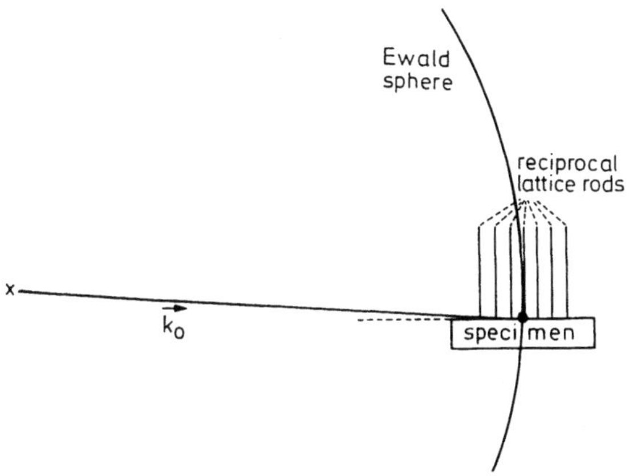

Fig. 10. Ewald construction for a RHEED pattern taken at glancing incidence from an extremely smooth monocrystalline GaAs film formed by MBE

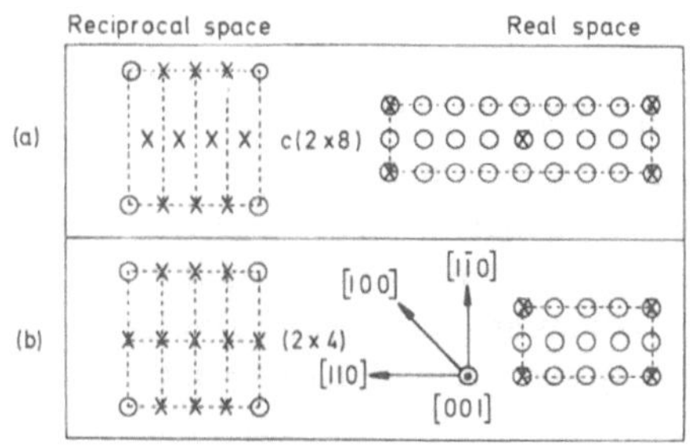

Fig. 11 a and b. Schematic representation of reconstructed C(2 x 8) and (2 x 4) surface structures in reciprocal and real space. Open circles represent "bulk" periodicities and crosses represent superlattice periodicities

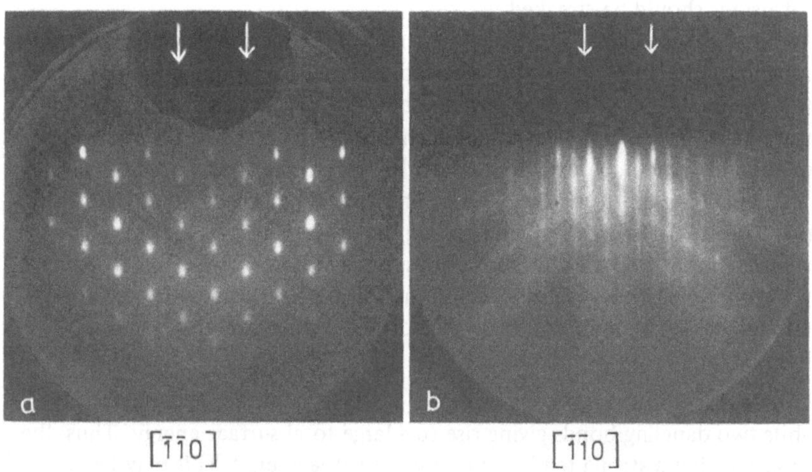

Fig. 12 a and b. 30 keV RHEED patterns of (100) GaAs surface taken at [1̄1̄0] azimuth; (**a**) substrate after chemical etching; (**b**) after deposition of 500 Å GaAs by MBE indicating strong reconstruction. The arrows show bulk diffraction features

temperature prior to deposition, results from a penetration of the incident electron beam through surface aspherities, indicating that the initial chemically etched surface is not atomically flat. The high background of the RHEED pattern is caused by an adsorbed amorphous surface layer due to the particular etching process applied. Cho[89] established that a substrate surface which was rough on an atomic scale quickly smoothed out after the deposition of only a few hundred Å of GaAs. This is indicated in Fig. 12b showing a RHEED pattern which has changed from spots to streaks normal to the shadow edge. Moreover, with the onset of growth, additional diffraction streaks due to surface reconstruction have occurred at fractional intervals between the elongated original spots. Then no further changes in the intensity modulation along the streaks appear, when the growth is continued under constant conditions. However, Neave and Joyce[37] have recently illustrated the ambiguities of streaked RHEED patterns used as a criterion to define the presence of a smooth surface with steps down to atomic dimensions. Although noted in many MBE laboratories for several years, Neave and Joyce were the first to provide definite experimental evidence from a combined RHEED-SEM study that even less smooth (100) GaAs surfaces may produce standard diffraction streaks with little intensity modulation normal to the shadow edge.

Therefore, a renewed attention is now focused on the presence and explanation of streaks in MBE-RHEED patterns. Bauer[88], Menadue[90], and Dove et al.[91] have proposed that streaks are observed if the surface contains disordered regions or imperfections of some nature. Alternatively Masud and Pendry[92] have suggested that a poorly collimated and monochromatized incident electron beam can result in streaks rather than in spots. While these interpretations are both related directly to the Ewald construction shown in Fig. 10, Holloway and Beeby[93] have recently pointed out that the occurrence of streaks in a RHEED intensity pattern is due to the influence of thermal diffuse scattering on the scattered beams. Consequently, any RHEED pattern which does not arise from transmission diffraction through surface protrusions should be streaked.

Aside from some ambiguity in correlating streaked diffraction patterns with the surface topography, RHEED observations have been extensively applied during MBE to determine the relationship between growth conditions and surface structure, and these measurements have often been complemented by AES and LEELS performed after completion of growth. Routinely the RHEED technique is now used for several purposes including (i) monitoring the cleaning and annealing process of the substrate surface prior to epitaxial growth; (ii) controlling the initial stages of the epitaxial growth; and (iii) monitoring the changes in the surface structure, when heterostructures are grown or when changes in the arrival rate of the constituent elements or of the substrate temperature are intentionally performed during growth.

2.2.5.2.1 Surface Reconstruction. On an ideal (100) GaAs surface each surface atom exhibits two dangling bonds giving rise to a large total surface energy. Thus, the ideal (100) surface has a strong tendency to lower its free energy. This may be accomplished by a variety of mechanisms, all leading to reconstructed surface structures. The process of reconstruction is a re-ordering of the arrangement of the outermost atomic layers at the surface of a crystal often leading to a surface structure having a periodicity different from the bulk lattice. This re-ordered phase is assumed to consist of fewer atoms than the ideal (bulk-terminated) surface.

A surface which reconstructs to a lower symmetry than the bulk crystal gives rise to extra diffraction lines in the RHEED pattern. A variety of surface reconstructions have been reported for the (100) GaAs surface during MBE growth[89,94,95]. The form of these structures depends critically on the incident fluxes of gallium and arsenic as well as on the surface temperature. The most relevant structures for practical film growth have been found to be the GaAs(100) – C(2 x 8) [(2 x 4)] and the GaAs (100) – C (8 x 2) [(4 x 2)] structure. The As-stabilized C (2 x 8) [(2 x 4)] structure is commonly produced during MBE growth, which is generally performed under As_4/Ga flux ratios in excess of 0.5. The Ga-stabilized C(8 x 2) [(4 x 2)] structure may be obtained by film growth under an As_4/Ga flux ratio of about 0.5 or by heating the As-stabilized structure to $T_s \geqslant 550$ °C.

The real-space and the reciprocal-lattice representation for the C(2 x 8) and (2 x 4) surface unit mesh are schematically depicted in Fig. 11. The corresponding diffraction patterns for the As-stabilized structure taken along the two most informative azimuthal directions are shown in Fig. 13. The diffraction patterns are represented by cuts of the smoothly curved Ewald sphere (cf. Fig. 11) through reciprocal space, i.e., in the present case normal to the plane of the figure and normal to the respective azimuths. More detailed representation of reciprocal lattice sections and associated schematic RHEED patterns for various reconstructions has been given by Cho et al.[1,89,96], Ludeke[97], and Neave and Joyce[37]. The diffraction patterns in Fig. 13 show streaks for both azimuths separated by a distance proportional to the inverse of the bulk atomic spacings. In addition, the pattern for the [$\bar{1}\bar{1}$0] direction exhibits 1/2 order and that for the [1$\bar{1}$0] direction 1/8 (or 1/4, resp.) order streaks as well, indicating the presence of a reconstructed surface structure with all spacings two and

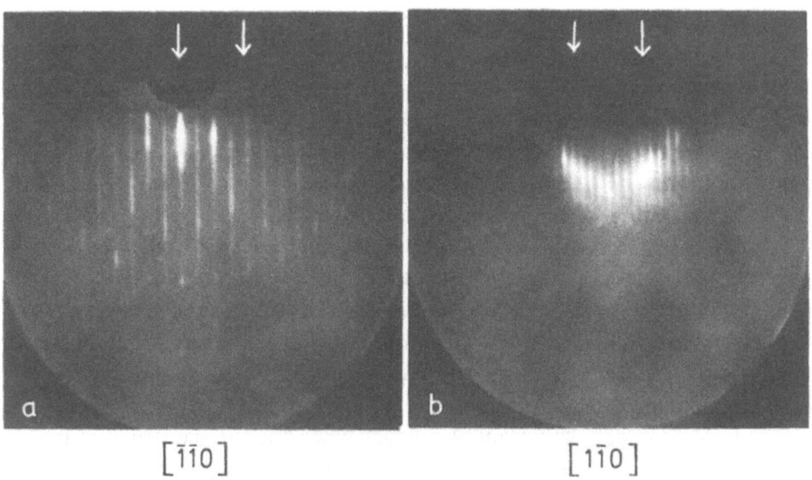

Fig. 13. 30-keV RHEED patterns of (100) GaAs film grown with As-stabilized C (2 x 8) or (2 x 4) surface reconstruction. Azimuths as shown. The arrows indicate location of bulk diffraction features

eight (or four, resp.) times as large as those of the bulk material. The Ga-stabilized C (8 x 2) and (4 x 2) structures exhibit the same RHEED patterns as the C (2 x 8) and (2 x 4) structures, but with the two [110] azimuths interchanged. This means that in real space the two structures are transformed into each other by a 90° rotation. In the RHEED patterns obtained from the two principle structures on (100) GaAs, distinct differences in the lengths of the streaks and in their intensity distributions have been observed. Generally, the patterns generated from the azimuth exhibiting the two-fold spacing show much longer streaks with very little intensity modulation compared to those of the eight-(or four-, resp.) fold spacing (cf. Fig. 13).

Neave and Joyce[37)] have recently pointed out that it is impossible to distinguish between the centered C(2 x 8) and the non-centered (2 x 4) structure or between the corresponding C(8 x 2) and (4 x 2) structures using RHEED. A distinction would require a diffraction pattern in a [110] azimuth including a 1/2 order Laue zone which, however, has never been observed in any published RHEED work on (100) GaAs. Therefore, many authors now refer simply to a (2 x 4) or (4 x 2) reconstruction. This does not imply, of course, that centered structures do in principal not occur but only that, unlike LEED, RHEED does not provide any distinct diffraction features associated with centered structures *during* MBE growth.

Depending on the incident arsenic and gallium fluxes and on the substrate temperature, reversible transitions between the two principal surface structures on (100) GaAs are possible, which are involved with a change of the surface composition. Several intermediate structures, e.g., (3 x 1), (1 x 6), (4 x 6), (3 x 6), and mixtures, can be observed within very narrow ranges of growth conditions. Whereas the transition between these structures are not sharp, the final change to the Ga-stabilized C(8 x 2) [(4 x 2)] structure is very abrupt. Furthermore, only a very small increase (~3%) in the Ga flux above the minimum value required to produce the Ga-stabilized structure results in the formation of free gallium on the surface. This means that the C(8 x 2) [(4 x 2)] reconstruction apparently represents a condition of maximum gallium surface population and that the stability range of this reconstruction with respect to the gallium flux is very small.

Using the flash desorption technique, Arthur[71)] has demonstrated that the surface structural changes on (100) GaAs are in fact correlated with gain or loss of arsenic from the surface since As atoms are only weakly bonded on the surface in comparison to Ga atoms. He suggested that for an As-stabilized structure the site occupancy of arsenic on the surface ranges from 50 to 60%, whereas for a Ga-stabilized structure less than 10% of the surface sites should be occupied by arsenic, i.e., the surface structural changes result from the gain or loss of ~0.5 monolayer of As atoms. With respect to the relationship between reconstruction and surface stoichiometry, attempts have been made by Cho[96)] and by Ploog and Fischer[20)] to measure at least qualitatively the relative Ga/As ratio on the surface for different surface structures by AES using fairly low current densities of the primary electron beam. Figure 14 shows how the $L_3M_4M_4$ peaks of Ga and As change when the surface structure is changed from Ga-C(8 x 2) to As-C(2 x 8). The ratio of $\frac{GaL_3M_4M_4}{AsL_3M_4M_4}$ peaks is 1.1 for Ga-C(8 x 2) and 0.8 for As-C(2 x 8). The $AsL_3M_4M_4$ peak increases by 30%. Since the escape depth of the high-energy Auger electrons is in the order of several mono-

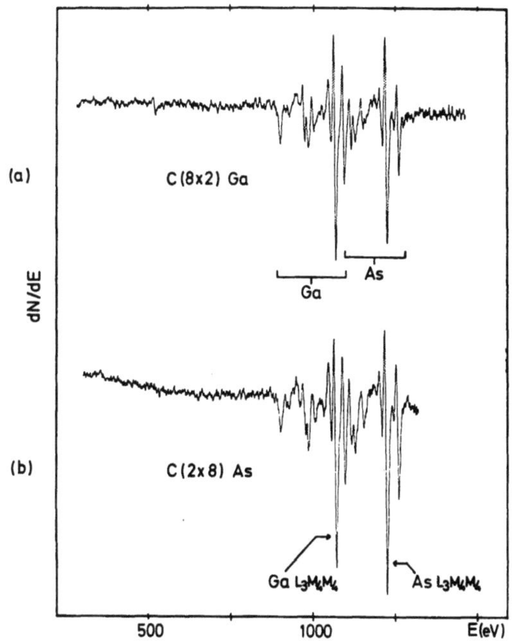

Fig. 14a and b. Auger spectra of Ga-stabilized C(8 x 2) or (4 x 2) and As-stabilized C(2 x 8) or (2 x 4) surface structures on (100) MBE GaAs indicating the change in Ga and As peak intensities, respectively

layers[76)], the Ga to As ratio on the surface cannot be measured very accurately by this method. The fairly large increase of the ratio indicates, however, that more than just the outermost atomic layer is involved in surface stoichiometry and thus the formation of reconstructed surface structures. This assumption is apparently supported by careful LEELS measurements performed by Joyce et al.[98)].

The relationship between reconstruction and surface stoichiometry has an important consequence on the incorporation of amphoteric dopants, such as Ge, during molecular beam epitaxy[99)]. When GaAs layers are grown with an As-stabilized C(2 x 8)[(2 x 4)] surface structure, Ge is incorporated principally as a donor, and as an acceptor while the layers are grown with a Ga-stabilized C(8 x 2)[(4 x 2)] surface structure. Thus p/n junctions in GaAs can be obtained by impinging molecular beams containing Ge, Ga, and As_4 simultaneously on the substrate and by modifying intentionally the As_4 to Ga ratio at a constant substrate temperature[100)]. The variation of the As_4 to Ga ratio, which yields different surface structures, can easily be monitored by the RHEED technique. Details of this method will be discussed in Sect. 2.3.2.

Neave and Joyce[37)] recently obtained an important result concerning maximum As surface population on (100) GaAs. The As-stabilized C(2 x 8) and (2 x 4) structures were found not to represent saturation with respect to arsenic. At a temperature ranging from 380 to 500 °C a C(4 x 4) reconstruction can occur which corresponds to a relatively higher As surface population than is present in the C(2 x 8) and (2 x 4) structures. The C(4 x 4) structure can also be maintained during growth at extremely high As_4 to Ga flux ratios ($J_{As_4}/J_{Ga} \gg 100$).

Surface potentials and charge densities for ideal and reconstructed surfaces of III–V compounds were calculated by Appelbaum et al.[101, 102)]. Based on these theoretical calculations, Cho[96] has experimentally determined the orientation of the surface structures of the As-stabilized (2 x 4) and C (2 x 8) and the Ga-stabilized (4 x 2) and C (8 x 2) structures on (100) GaAs with respect to the atomic bond direction. Using RHEED he found that for either stabilization the two-fold periodicity of the surface structures has a direction parallel to the plane containing the dangling bonds in the surface and perpendicular to the plane containing the backbonds. This analysis was independently confirmed by van Bommel and Crombeen[103] using LEED.

A more detailed reconstruction model and any physical impetus for reconstruction, however, would be highly speculative at this time, because of the extreme complexity of the (100) GaAs surface. The latest investigations of Neave and Joyce[37] would require reconstruction to be related to vacancies according to a model suggested by Phillips[104] rather than being derived from Fermi-surface instabilities. Hopefully, some recent studies by Ludeke and Ley[105] using photoelectron spectroscopy might bring about some new ideas to help solve the apparent complexity of surface structures on (100) GaAs. The authors found that the valence electron spectra of both the (110) and the (100) GaAs surfaces revealed six sharp peaks, which were associated with critical points in the density of surface states. The similarity in the spectra of the (110) and (100) surfaces suggests that the (100) surface consists of (110)-like micro facets, which are smaller than the dimensions of the commonly observed Ga- or As-stabilized reconstructions on the (100) GaAs surface.

The most important experimental feature of RHEED is that it enables the operator to monitor growth continuously from the onset. The diffraction patterns generated by different surface structures are highly reproducible for given beam fluxes and substrate temperatures. Once a well-developed streaked diffraction pattern indicating an As-stabilized (2 x 4) or C (2 x 8) reconstruction has been produced (Fig. 13), then epitaxial growth starts with an undistorted continuation of this structure. Similar surface structures as described for (100) GaAs have been found for many other III–V compounds and also throughout the alloy range.

Characteristic differences in the intensity modulation along the streaks and in the stability ranges of various structures do, however, occur. This makes it feasible to correlate the growth process with the lattice mismatch over the entire composition range during heteroepitaxy of III–V alloy films on various substrates[106, 107].

2.2.5.3 Surface Studies by Secondary Ion Mass Spectroscopy (SIMS)

The application of secondary ion mass spectroscopy (SIMS) for pre- and post-deposition analysis during MBE has been stimulated by the fact that a sensitive quadrupole mass analyser for residual gas analysis and an Ar^+-ion gun for substrate cleaning are in any case incorporated into many MBE apparati. Without any detraction from these assignments, ion gun and mass filter can be adjusted in an appropriate geometry as to also perform SIMS experiments within the growth chamber[20].

In order to reduce the damaging effect of the exciting primary beam, the sample surface is bombarded with only a small primary ion dose. In this "static mode"[108], SIMS is a powerful tool for determination of the chemical composition of the outermost atomic layers of a solid. Compared to AES, the main advantages of SIMS are high absolute sensitivity for many elements, given an ample supply, good elemental specifity and high mass resolution. When high erosion rates of about 10 Å/s are chosen by applying high primary ion currents, the SIMS technique can also be used for analysis of bulk material and for surface profiling.

With the improvements in the primary ion source, in the secondary ion optics and in the ion detection, SIMS has recently become a true surface and a powerful depth profiling technique by removing particles from the solid, layer by layer. Even the analysis of surface impurities can be accomplished via the ion bombarding technique. However, for compound semiconductors there are a number of rather complicated and as yet incompletely understood processes, which are responsible for the emission of positive or negative ions. For this reason there are as yet very few examples of an application of the SIMS technique during MBE growth of semiconductor materials.

Ploog and Fischer[79)] studied the surface accumulation of the dopant Sn during MBE growth of GaAs. These SIMS investigations of Sn-doped GaAs layers prepared under different growth conditons and with different dopant concentrations yielded results comparable to those found by AES (see Sect. 2.3.2). For carrier concentrations $7 \times 10^{16} \leqslant n \leqslant 1 \times 10^{18}$ cm^{-3} the amount of Sn which remains on the growing surface during MBE of Sn-doped GaAs is about three orders of magnitude larger than that which is incorporated into the growing film. The thickness of the Sn-enriched surface layer can be estimated to be in the range from 10 to 20 Å.

Dowsett et al.[109)] have evaluated the residual contamination on (100) InP substrates. Even after ion-etching contamination levels in the 100 ppm range about 0.48 μm from the surface have been found to persist. Such high residual contamination levels on substrate surfaces may adversely affect nucleation and defect incorporation in the growing films. Therefore, the SIMS technique with its high surface sensitivity may be usefully employed to monitor substrate preparation procedures aimed at reducing the contamination of the initial surface. If these contamination levels, which are below the detection limit of AES, can be further reduced, then some general improvements in the structural and electrical properties of the films could be anticipated.

2.2.5.4 Studies of Structural and Electronic Surface States

Most of the investigations of structural and electronic surface states of compound semiconductors have until recently been performed on surfaces which can easily be prepared in a clean and reasonably ordered condition by appropriate vacuum cleavage techniques. For GaAs this is the (110) face. Now, for a few years, molecular beam epitaxy has enabled us to prepare flat, well-ordered, clean surfaces, the composition of which can be altered in a reproducible manner. Thus, an increasing number of well-defined semiconductor surfaces with a variety of orientations have now been made readily available for studying the electronic surface states with more sophisticated techniques. The most versatile and inexpensive method to study the electronic properties of the outermost atomic layers of MBE grown material has turned out to be *low-energy electron loss spectroscopy* (LEELS). This is because the same analyser system (CMA) and geometry can be used for both Auger and LEELS measurements. In the field of *photoelectron spectroscopy* (UPS and XPS) there is considerable activity, as the new spectroscopic photoemission partial-yield technique (using a continuously tunable source of radiation) for probing empty surface states and conduction band features promises to enhance our understanding and control of technologically interesting metal-semiconductor Schottky barriers[110].

Recently, Larsen et al.[111] first combined the technique of angular resolved photoemission spectroscopy with MBE and RHEED to study effects of reconstruction on the electronic properties of (100) GaAs surfaces. The authors observed two prominent surface states near the top of the valence band maximum on both the As-stabilized (2 x 4) and Ga-stabilized (4 x 2) reconstructed (100) GaAs surfaces grown by MBE. The reported experimental data indicate that neither surface is saturated with As or Ga, respectively. In addition, Larsen et al.[111] showed that the symmetry requirements of the repeated surface Brillouin zone for the reconstructed (2 x 4) As-stabilized surface are obeyed for the As-associated surface state.

Additional information on semiconductor surface states at interfaces (e.g., vacuum-semiconductor, metal-semiconductor, and insulator-semiconductor) can be obtained by *Raman scattering* using laser excitation lines with different penetration depth into the semiconductor material. When applied to (100) oriented n-GaAs with high donor concentrations ($n \geqslant 2 \times 10^{18}\ cm^{-3}$) in a backscattering geometry, this technique yields both the coupled LO phonon-plasmon modes from the bulk and the unscreened LO phonons from the surface depletion region[112]. The relative intensities can be used to study the surface band bending and the barrier heights of free and covered GaAs surfaces prepared in situ by MBE techniques[113].

The few representative examples may be sufficient to show that molecular beam epitaxy is not only a versatile technique for the fabrication of special semiconductor materials for devices, but also a promising method for in situ preparation of well-defined surfaces for more thorough surface investigations, which might be performed in the near future.

Finally, a careful reflection on surface studies during MBE must emphasize the fact that not all of this rather expensive equipment and not all of the time associated with these experiments is required for *every* MBE application, in order to produce high quality compound semiconductor films in a reproducible manner. Once the optimal growth parameters have been evaluated in a more sophisticated growth apparatus, like the one described before, the data can be transferred to similar yet simple UHV apparatus in the same laboratory designed for the growth of materials for optoelectronic device technology. An MBE apparatus for this purpose has recently been introduced by Barnes and Cho[114] and is schematically shown in Fig. 15. An electron gun for RHEED is the only surface diagnostic technique installed. The substrate which is inserted by a sample exchange system is entirely surrounded by a special cryopanel. The extensive use of these liquid-nitrogen-cooled shrouds drastically reduces the CO and H_2O background and any further contamination due to outgassing from the chamber walls.

Most of the integral MBE systems which are now commercially available from several manufacturers[115] consist of a growth station similar to the one shown in Fig. 15. This growth chamber is designed to be directly connected to a separate surface analysis chamber via a special isolation valve and sample transfer system.

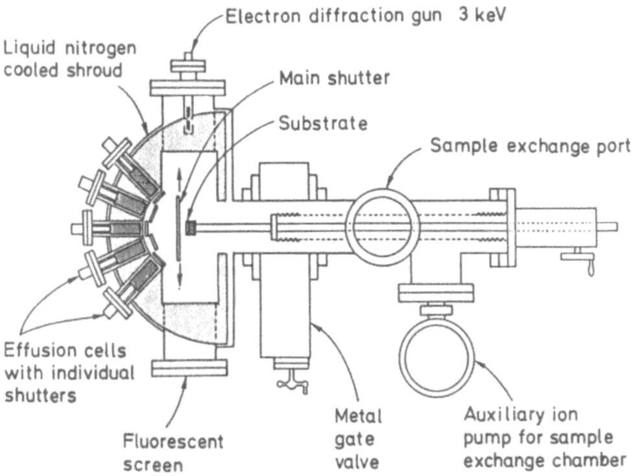

Fig. 15. Schematic diagram of a newly designed MBE apparatus with extensive use of cryopanels in the growth chamber (from Ref. [114])

2.3 Growth Mechanism and Dopant Incorporation

Crystal growth during molecular beam epitaxy is governed predominantly by kinetics as opposed to the quasi-equilibrium conditions existing during LPE or VPE growth Nevertheless, our insight into the growth mechanism – while still incomplete – is probably best for MBE, since the results obtained by an application of the multiple in situ methods of surface characterization described in Sect. 2.2.5 and the results of surface kinetic measurements to be discussed have contributed much to the establishment of a growth model. Although most of the studies have been made on GaAs, the general growth process is probably not markedly different for other III–V compounds.

For dopant incorporation during molecular beam epitaxy a general mechanism does not exist so that each potential dopant element must be tested empirically. A detailed model for the nonequilibrium incorporation of donor impurities into electrically active sites has been established only for Sn[22]. However, there are now sufficient empirical data available for several other elements suitable as electrically active impurities for MBE that both high-quality n- and p-type GaAs can be grown. The behaviour of these elements which have been used successfully will be discussed in Sect. 2.3.2.

2.3.1 Growth Mechanism

The earliest growth model for compound semiconductors was established by Günther[116]. He proposed that the concentration of III–V nuclei should be proportional to the product of the concentrations of adsorbed group-III and group-V elements. In practice, however, this model failed because it was based simply on thermodynamic considerations.

The currently accepted growth model has been developed from studies of growth rates, surface morphologies and reaction kinetics. The surface morphology of MBE GaAs films with respect to growth mechanism was first studied by Cho et al.[1, 89, 117] using combined RMEED and electron microscopy of surface replicas. The observation of the substrate surface smoothing process (Fig. 12) indicates that the growth mechanism is via a two-dimensional step propagation. Surface aspherities should be excellent step sources. Undesired three-dimensional nucleation is apparently always associated with a high surface impurity concentration (Fig. 9) or with mechanical damage. The source of growth steps on atomically flat surfaces, however, still remains speculative. Rode[118] and van Vechten[119] have suggested that there is a direct relation between surface reconstruction and surface steps. The atoms set free by the reconstruction process should move across the surface in a front until they encounter another such wave to produce the next growth step. However, this idea cannot explain the two-dimensional growth without reconstruction which occurs at growth temperatures below ~300 °C[64].

Surface kinetic data on growth and doping processes have been obtained by modulated molecular beam techniques. These experiments were initiated by Arthur[4], and they have been continued and improved by Foxon et al.[27, 72–74, 120]. The experimental setup for modulated molecular beam mass spectrometric studies used by Foxon et al. is schematically shown in Fig. 16. The effusion cells installed[27, 33] can produce neutral beams of atoms or molecules having thermal velocities. The beam intensities range from 10^{10} to 10^{15} molecules/cm^2 · s. Either the incident flux or the desorbing flux is periodically modulated. In either case the flux desorbing from the substrate is detected in a sensitive quadrupole mass spectrometer, the ioniser of which is completely surrounded by a cryopanel. Both phase sensitive detection and a

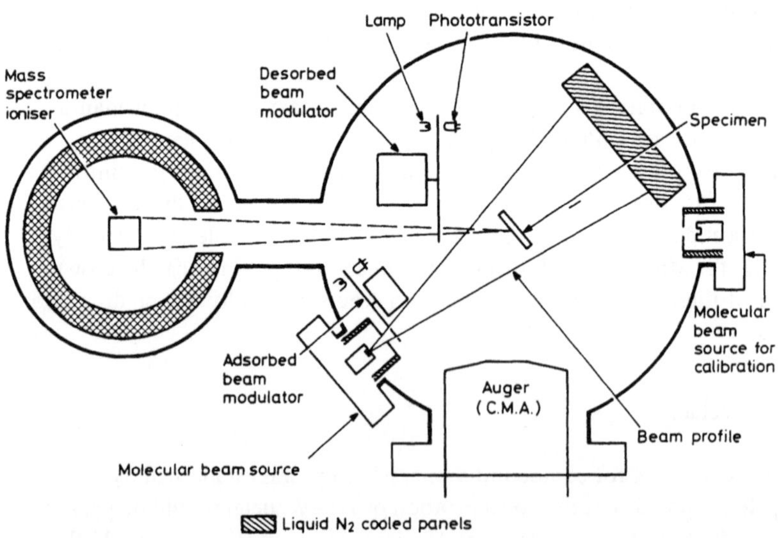

Fig. 16. Experimental arrangement for modulated molecular beam studies, as used by Foxon and Joyce[27, 72–74, 120]

real-time, signal-averaging system have been used[27] in order to separate beam from background signals in the mass filter. The data were processed by time-domain analyses together with Fourier-transform techniques[27]. For a detailed assessment of the substrate surface, AES and – not shown in the Figure – RMEED facilities have also been incorporated into the apparatus. The kinetic data which can be extracted from modulated molecular beam experiments include thermal accommodation coefficient, surface residence time, desorption energy, sticking coefficient, and order of reaction of the impinging species. From these data the authors have deduced detailed mechanisms for the interaction of gallium, arsenic, and some dopants with (100) GaAs surfaces.

Using a pulsed molecular beam technique, Arthur[4, 71] has observed that below 480 °C Ga has a unity sticking coefficient on (100) GaAs. Above this temperature Ga is adsorbed as well as desorbed. The behaviour of As_2 (generated from a GaAs source), however, is more complex. A condensation of As_2 occurs only when Ga

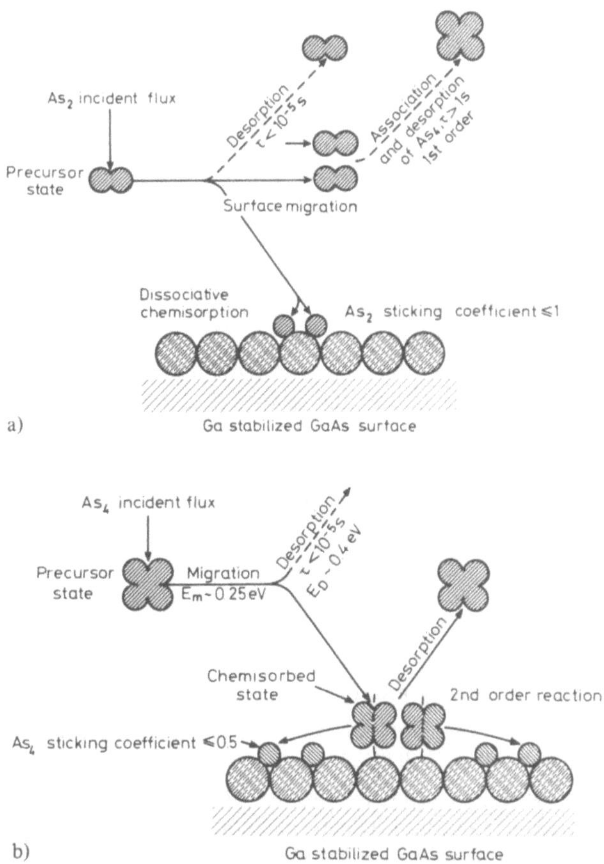

Fig. 17a and b. Models for the interaction of gallium and arsenic molecular beams on a (100) GaAs surface resulting in epitaxial growth; (a) molecular beams of Ga and As_2; (b) molecular beams of Ga and As_4 (from Ref. [73, 74, 121])

adatoms are supplied from an extra Ga beam or when free Ga is created on the GaAs surface by non-congruent surface dissociation at $T_s \geqslant 500\,°C$. Joyce et al.[74, 121] deduced a detailed growth model for GaAs from Ga and As_2 molecular beams which is schematically shown in Fig. 17a. According to this model As_2 molecules are first adsorbed into a mobile, weakly bound precursor state. Dissociation of adsorbed As_2 can only occur when the molecules encounter paired Ga lattice sites while migrating on the surface. In the absence of free surface Ga adatoms, As_2 has a measurable surface lifetime but no permanent condensation occurs. The sticking coefficient of As_2, S_{As_2}, is thus a function of the arrival rate of Ga, J_{Ga}, being unity only for a monolayer of Ga atoms or for $2\,J_{Ga} \geqslant J_{As_2}$. At substrate temperatures below 330 °C a pairwise association of adsorbed As_2 molecules followed by subsequent desorption as As_4 molecules commences and becomes increasingly dominant[74]. From these kinetic measurements it follows that stoichiometric GaAs can be grown for relative flux rates of As_2 and Ga larger than or equal to unity.

A similar growth model, schematically shown in Fig. 17b, holds for the interaction of Ga and As_4 molecular beams with a heated (100) GaAs surface[72, 121]. Again, the As_4 molecules (generated from an elemental arsenic source) are first adsorbed into a mobile precursor state and the Ga adatom population controls the condensation and reaction of As_4 at $T_s \geqslant 180\,°C$. The crucial result reported by Foxon and Joyce[72] is that the sticking coefficient of As_4, S_{As_4}, can never exceed the value of 0.5, even when the GaAs surface is covered by a monolayer of Ga atoms. This behaviour has been explained by a pairwise interaction process of As_4 molecules chemisorbed on adjacent Ga lattice sites, which is involved in the surface chemical reaction. As a rough guideline the GaAs films grow with an As-stabilized C(2 x 8) or (2 x 4) surface structure when $J_{As_4} \geqslant 2\,J_{Ga}$, i.e., $S_{As_4} \leqslant 0.5$; and the Ga-stabilized C(8 x 2) or (4 x 2) structure is observed, when $J_{As_4} \leqslant 2\,J_{Ga}$, i.e., $S_{As_4} \approx 0.5$. The exact value of the flux ratio depends on the absolute Ga flux[37]. For practical film growth the important result from the kinetic studies is that stoichiometric GaAs can be grown for a whole range of substrate temperatures and relative flux ratios of As_4 and Ga larger than or equal to 0.5.

The growth rate R (µm/h) of MBE GaAs films is entirely controlled by the flux density of the Ga beam, F (atoms/$cm^2 \cdot s$), impinging on the substrate surface[1, 11], i.e.,

$$R = \alpha \cdot F \tag{2}$$

where α is the sticking coefficient of Ga which has a value close to unity at typical MBE growth temperatures (450–620 °C) under As-stabilized growth conditions. This implies that nearly all the Ga atoms incident on the surface get incorporated into the growing epitaxial layer. It should be noticed, however, that these Ga atoms, after being adsorbed on the substrate surface, still undergo a considerable amount of migration and rearrangement. The validity of Eq. (2) has been determined by measuring the growth rate from micrographs of the film thickness and comparing it with the Ga flux which can now be measured with an overall accuracy of ±8% using a calibrated ion gauge[37].

The growth characteristic of MBE over small substrate areas with dimensions on the order of the surface diffusion length, however, is quite different from that over large areas. For growth on small areas, Tsang and Cho[122] have shown that the arriving atoms may diffuse to an adjacent facet if the growth on that plane is more favorable. To account for the surface diffusion effect, they have therefore modified Eq. (2) and introduced another parameter β, i.e.,

$$R = \alpha \cdot \beta \cdot F = \gamma \cdot F \tag{3}$$

The product $\alpha \cdot \beta$ was denoted the effective sticking coefficient of a facet, γ. The quantity β is unity for large surface areas. When the surface dimension is in the order of the diffusion length, $\beta \leqslant 1$ if growth on the adjacent facet is more favorable and $\beta \geqslant 1$ if growth on the adjacent facets is less favorable. From measured layer thicknesses, Tsang and Cho[122] have obtained approximate values of $\beta \approx 1$ for (100) planes and $\beta \approx 0.5$ for (111) planes, i.e., MBE growth of GaAs on (100) planes is more favored than on (111) planes. The growth characteristics with MBE thus turned out to be entirely different from those obtained by similar LPE experiments. With respect to this effect, Tsang and Cho[122] have suggested that MBE growth of GaAs/$Al_xGa_{1-x}As$ multilayer structures over preferentially etched channels on GaAs substrates may be particularly suitable for fabricating electrooptical devices which require planar definition and for monolithic integration (see Sect. 4.1.3).

The temperature range for MBE growth of monocrystalline GaAs films is apparently rather wide. However, the minimum temperatures quoted by various authors[18, 89, 95] differ considerably. The reason for these differences is probably the status of the substrate surface on which growth commences. A consideration of the surface chemistry involved in the growth of GaAs from Ga and As_4 molecular beams on (100) GaAs led Foxon and Joyce[72] to the conclusion that GaAs should be formed at substrate temperatures at least as low as 180 °C, since the adsorption of As_4 molecules becomes non-dissociative at $T_s \leqslant 180$ °C.

Recently, Neave and Joyce[64] have made a systematic investigation of substrate temperature effects on MBE growth of GaAs using RHEED. They established a surprisingly large temperature range for monocrystalline growth extending to particularly low temperature. Provided the starting substrate surface is sufficiently clean and perfect, monocrystalline and smooth GaAs films can be grown autoepitaxially at substrate temperatures ranging from 630 °C to as low as 90 °C with reasonable growth rates of 0.5–1.0 µm/h. From 600 to 260 °C the growing GaAs films maintain a reconstructed surface structure, while from 260 to 90 °C a diffraction pattern arising from the bulk surface structure only is observable. However, above and below the temperature at which reconstruction disappears the films remain monocrystalline and extremely smooth. Using electron channeling, Neave and Joyce[64] showed that the crystal perfection of a GaAs film grown at 100 °C is comparable to a film grown at 520 °C. There is apparently no major change of growth mode associated with the loss of surface reconstruction, although some as yet undetermined features have recently been observed by transmission electron microscopy[123] and keep the subject a matter of speculation. From 90 °C to 70 °C a transition range in the growth mode occurs and below 70 °C merely amorphous material is deposited. The high temperature limit for controlled MBE growth is given by the congruent evaporation of GaAs which occurs at 625 to 635 °C[38, 47].

While the growing GaAs films remain monocrystalline to substrate temperatures as low as 90 °C, the electrical and optical properties of the films deteriorate drastically when grown at $T_s \leqslant 420$ °C. GaAs films grown below 400 °C are highly compensated and exhibit a low mobility as well as very poor photoluminescence properties. In practise, therefore, GaAs films can be grown by MBE at substrate temperatures ranging from 450 to 630 °C. Tsang[124] has recently provided some evidence that photoluminescence is improved when the epilayers are grown at $T_s \geqslant 600$ °C. In order to ensure stoichiometric growth, an arsenic flux density in excess to that of gallium must be applied. Generally, the GaAs films are grown with an As-stabilized surface reconstruction, because the thus required growth conditions are easier to maintain.

Another interesting technique for the study of crystal growth processes on an atomic scale has been proposed by Petroff et al.[41]. This technique comprises epitaxial growth of alternate monolayer structures of GaAs and AlAs (see Sect. 4.2) followed by examination of cross-sections of the crystals with the various imaging techniques allowed by transmission electron diffraction. The authors reported the effects of both chemical composition and growth temperature T_S on the growth of $(GaAs)_n$-$(AlAs)_m$ superlattices on (100) GaAs. Critical temperatures, $(T_S)_C$, above which surface roughening effects prevent the formation of an ordered layered structure were found. $(T_S)_C$ is a function of the chemical composition of the film. The results of Petroff et al.[41] point out the importance of the growth parameters on the quality of the interfaces in MBE grown superlattices (see Sect. 4.2).

The growth model established from kinetic measurements is not unique to GaAs but also valid for AlAs[125], InP[70], for a number of other III–V compounds, and – with minor modifications – also for ternary III–III–V alloys. In practical terms, $Al_xGa_{1-x}As$ films having good crystallographic perfection can be grown by simply directing beams of Ga, Al, and excess As_4 simultaneously to the growth interface. Since the sticking coefficient of Al on GaAs is unity[125], a good compositional control of the growing film can be achieved by adjusting the flux densities of the impinging group III beams. Recently, Foxon and Joyce[19] found that a transfer of the growth parameters previously determined for GaAs to the $Al_xGa_{1-x}As$ system had worked without complications concerning the compositional uniformity and control only, because AlAs happens to be more thermally stable than GaAs. From a thorough kinetic study of the involved surface processes they have deduced that the principal limitation to the growth of III–III–V alloys by MBE is the thermal stability of the less stable of the two III–V compounds of which the alloy may be considered to be composed. At higher temperatures preferential desorption of the more volatile group III element occurs. Thus, the surface composition of the alloy reflects the relative flux ratio of the group III elements only, if growth is carried out at temperatures below which GaAs in the case of $Al_xGa_{1-x}As$, or InAs in the case of $Ga_xIn_{1-x}As$, or InP in the case of $Ga_xIn_{1-x}P$ are thermally stable.

The growth of III–V–V alloy films by MBE is apparently much more complex since the relative amounts of the group V element incorporated into the growing film are not simply proportional to their relative arrival rates. This behaviour makes it extremely difficult to obtain a good compositional uniformity and control for MBE growth of these materials.

2.3.2 Dopant Incorporation

The incorporation of controlled amounts of electrically active impurities into the growing film is generally achieved by using additional effusion cells which contain the appropriate doping element. The resulting electrical and optical properties of the grown films depend basically on two processes. First, they rely on the flux of the impurity atoms reaching the growth interface, on their sticking probabilities and on their surface lifetimes, particularly in the presence of large fluxes of gallium and arsenic. The second process is the actual incorporation behaviour (e.g., possibility of displacement reactions leading to a perturbation of growth morphology) and the extent of electrical activity of the dopant. Some of the electrical and optical features of doped films which are closely related to film properties will be discussed in Sect. 3.

Direct kinetic measurements using modulated-beam techniques have been made for the potential dopant elements Zn, Cd, Be, Mg, Mn, Ge, Sn, Pb, and Te with GaAs as the substrate. All elements, except Zn, which have been used successfully as dopants for MBE GaAs have unity sticking coefficients over the whole range of growth conditions; i.e., substrate temperature up to 630 °C, impurity flux 10^9 – 10^{14} atoms/cm^2 · s, As_4/Ga flux ratio varied widely with an absolute intensity up to 10^5 greater than that of the dopant. This behaviour is indicated by Clausius-Clapeyron-type plots which have been established for most suitable doping elements. The Figs. 18 and 21 show two typical examples of measured and calculated free carrier concentrations plotted as function of the corresponding effusion cell temperatures.

At the present time four suitable *acceptor* impurities have been found for MBE GaAs. These are summarized in Table 1 along with some of their characteristic properties. The most recently established acceptor is *beryllium* which behaves as an almost ideal p-type dopant in MBE GaAs[30, 121)]. There are no complications during Be incorporation arising from segregation or anomalous diffusion effects. Figure 18 shows that the calculated and measured values of free acceptor concentration are directly

Table 1. Promising p-type dopants in MBE GaAs

Element	Incorporation behaviour	Degree of compensation	Maximum achievable hole concentration	Remarks
Be	Acceptor only	Low	5×10^{19} cm^{-3}	Vapor is highly toxic
Mn	Acceptor only	Low	1×10^{18} cm^{-3}	Film perfection deteriorates at $p \geqslant 10^{18}$ cm^{-3}, deep acceptor level
Ge	Depends strongly on growth conditions	Can be high	5×10^{18} cm^{-3}	Acceptor at low As_4 to Ga ratio and high growth temperature
Zn^+	Acceptor only	Low	2×10^{19} cm^{-3}	Sophisticated technique required for low energy ion production

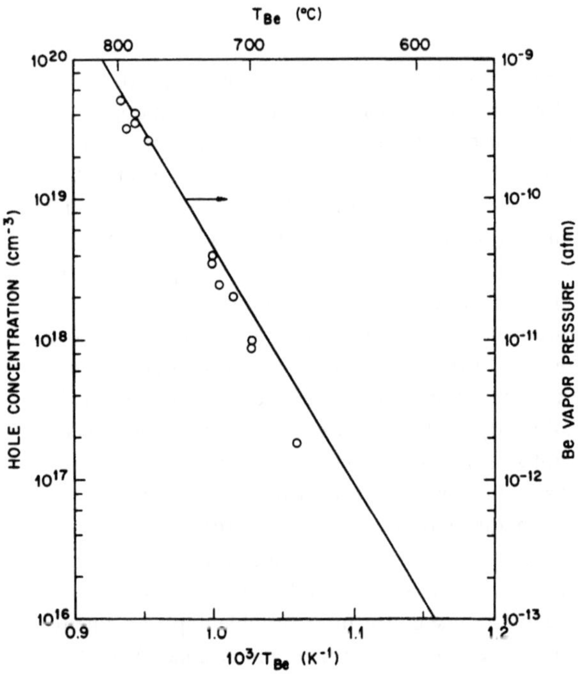

Fig. 18. Hole concentration as a function of reciprocal temperature of the Be source for p-type GaAs layers grown in the MBE system of Ilegems[30]. The solid line represents the Be equilibrium vapor pressure curve. The deviation from the straight line is probably due to the omitted factor of $T^{1/2}$ on the ordinate

proportional to the Be effusion cell temperature, i.e., Be vapor pressure. Each incident Be atom produces one free acceptor, which provides a fairly shallow acceptor level in GaAs only 30 meV above the valence band edge. Ilegems[30] has shown that measured doping profiles are flat and follow very closely ideal profiles calculated from the dopant beam intensity. He has demonstrated the ideal behaviour of Be with a typical dopant profile obtained from an n^+–p–p^+ diode structure in which the p-layer consists of a series of sharp Be-dopant pulses superimposed on a uniform Be-dopant background (Fig. 19). The measured "as-grown" profile matches almost exactly the theoretical free-carrier profile (top solid curve)[126, 127]. Little change was observed even after annealing at 800 °C for 1 h under an As_4 pressure of about 10 atm. Recently, McLevige et al.[128] found that the diffusion coefficient of Be at 900 °C in highly doped MBE GaAs ($p = 2 - 3 \times 10^{19}$ cm^{-3}) has a value of only $0.5 - 1.0 \times 10^{-10}$ cm^2/s. This value is extremely low, two orders of magnitude lower than that observed for implanted Be in GaAs of equal concentration. Using MBE, Be has thus turned out to be a very promising shallow acceptor for GaAs and also for $Al_xGa_{1-x}As$ thin films[30].

Before Be had been established to be a suitable acceptor, *manganese* was widely used for MBE growth of GaAs with As-stabilized growth conditions[29, 129]. However, the strong interaction of Mn with Ga on the growth surface resulted in a deteriora-

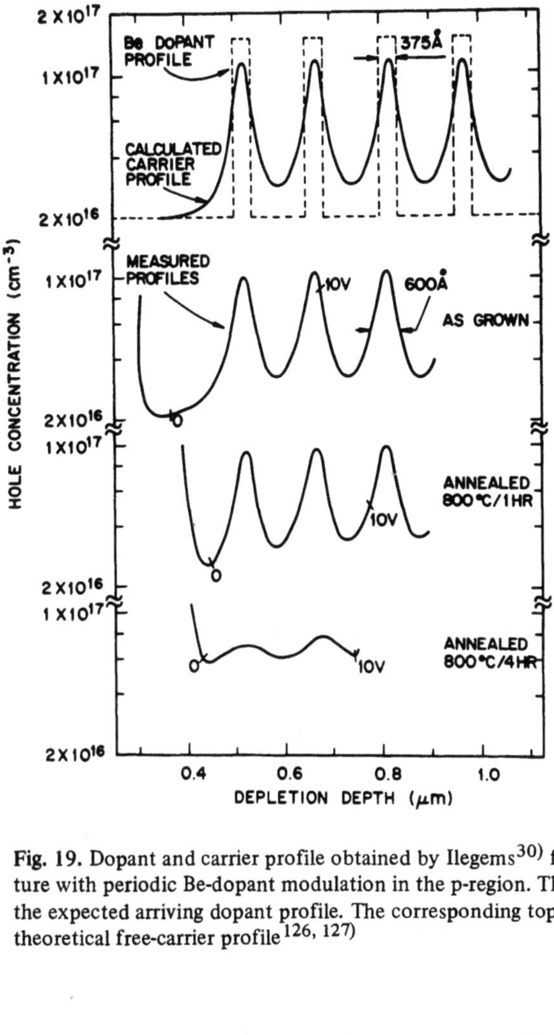

Fig. 19. Dopant and carrier profile obtained by Ilegems[30)] from a n^+-p-p^+ GaAs diode structure with periodic Be-dopant modulation in the p-region. The dashed line on the top represents the expected arriving dopant profile. The corresponding top solid line represents the calculated theoretical free-carrier profile[126, 127)]

tion of a film perfection with the three-dimensional island growth starting at acceptor levels $\geqslant 1 \times 10^{18}$ cm^{-3}. Moreover, Mn forms a rather deep acceptor (113 meV above the valence band edge) and is thus not fully ionized at room temperature.

The group-IV element *germanium* is amphoteric in epitaxial grown GaAs. For MBE GaAs this was first reported by Cho and Hayashi[99)]. On As sites Ge acts as an acceptor, while on Ga sites Ge acts as a donor. The site incorporation of the impurity element during growth depends thus critically on the As_4/Ga flux ratio and on the substrate temperature. This means that p/n junctions in GaAs can be grown by simply changing the growth conditions in an appropriate manner[99)]. When the surface of the growing GaAs film exhibits a Ga-stabilized C(8 x 2) or (4 x 2) reconstruction, the Ga atom surface population is increased. The codeposited Ge atoms are incorporated predominantly on As sites, and thus p-doped GaAs films result. The major problem with MBE growth of p-type GaAs films doped with Ge, which exhibit a low level of compensation and reasonable mobilities, has been the very small stability range of the required Ga-stabilized growth conditions with respect to the Ga flux while all other conditions remain constant. For practical purposes it has not been possible to maintain a Ga-stabilized C(8 x 2)[(4 x 2)] surface structure during continuous growth, resulting in a smooth mirrorlike surface appearance similar to that achieved routinely under As-rich growth conditions. Only a small increase (~3%) in the Ga flux above the minimum value required to produce the Ga-stable C(8 x 2) [(4 x 2)] reconstruction results in the formation of free gallium on the surface. Thus, until recently, most of the electrical and optical characterizations of MBE grown p-type GaAs films doped with Ge have been performed on samples with rough surfaces (see for example Ref. 30).

Improved reaction conditions have recently been established which enable the growth of Ge-doped p-type GaAs films with extremely smooth surfaces[61, 100)]. Cho[96)] and Neave and Joyce[37)] had shown that the concentration of As vacancies on the growth surface depends not only on the incident As_4/Ga flux ratio but also on the substrate temperature. Therefore, sufficient As vacancies can also be created, if the substrate temperature is increased to $T_s \geqslant 600\,°C$, while the As_4/Ga flux ratio is kept slightly above the value required to maintain the As-stabilized C (2 x 8) [(2 x 4)] reconstruction at 550 °C, i.e. $J_{As_4}/J_{Ga} \approx 2$. Under these growth conditions Ge is principally incorporated on As sites without an onset of three-dimensional growth, and the resulting p-type GaAs films exhibit reasonable mobilities (for details see Sect. 3). The elevation of the substrate temperature from $T_s = 550\,°C$ to $T_s \geqslant 600\,°C$ is accompanied by an overall change of the As-stabilized C(2 x 8)[(2 x 4)] surface structure to a (4 x 2) and finally a (1 x 1) structure without any reconstruction. The conversion of the surface structure can so be used to monitor the site incorporation of the incident Ge atoms in the RHEED pattern.

Although *zinc* has been used successfully as a p-type dopant in GaAs films prepared by LPE and VPE, during MBE it shows an extremely complex surface interaction behaviour on (100) GaAs[130, 131)]. The only well-established parameter is that at substrate temperatures ⩾ 100 °C the sticking coefficient of both neutral Zn and Cd is zero. Therefore, ionized atoms have to be used for an effective acceptor incorporation during MBE growth as indicated in Table 1. Takahashi et al.[132, 133)] have shown that an effective sticking coefficient of 0.01–0.03 on GaAs and $Al_xGa_{1-x}As$ can be achieved by using medium energy (0.2–1.5 keV) Zn^+ ions. However, Bean[134)] found rather poor minority-carrier properties in the as-grown Zn^+ doped GaAs films, and reasonable hole mobilities could be achieved only after annealing the samples at ~800 °C under As_4 pressure. In future, the application of low-energy ion beams might provide a very attractive doping technique in order to write in situ three-dimensional doping structures in GaAs during MBE growth.

Magnesium has a sticking coefficient close to unity on GaAs and $Al_xGa_{1-x}As$[1, 60)], but it has apparently a very low level of electrical activity[9, 58)]. Only *one* incident Mg atom from 10^5 seems to be active in GaAs. In $Al_xGa_{1-x}As$ the active acceptor concentration was found[58)] to increase rapidly with increasing Al content. The reason for this behaviour is not known. It has been speculated[9, 26)] that this low level of activity could be understood if Mg occupied mainly interstitial sites under such non equilibrium growth conditions. This idea might be supported by the anomalously high diffusion rate at which Mg was found to be transferred from the growth surface to the bulk[26)], possibly via those interstitial sites. Because of these uncertainties, Mg was used as a p-type dopant for GaAs and $Al_xGa_{1-x}As$ only in early MBE work. Net acceptor concentrations as high as 10^{19} cm^{-3} have been achieved[58)].

Until recently the incorporation of *donor* impurities in MBE GaAs seemed to be less complicated as compared to acceptor incorporation. The user has now the choice of three (or possibly six) donor impurities for MBE of GaAs, which are summarized in Table 2 along with some of their characteristic features. Apparently, however, none of them is without unfavorable attributes. For practical applications, n-type doping of III–V compounds has been limited mainly to the group IV elements Si, Ge, and Sn which are, however, potential amphoteric dopants. As yet, only few studies[1, 71, 135, 136)] of n-type doping of GaAs with group VI elements have been performed.

The most convenient n-type dopant for GaAs in any epitaxial growth technique is *tin*. Contrary to the other group IV elements Si and Ge which are amphoteric and can act as donors or acceptors in GaAs depending on the growth condition, with Sn only n-type GaAs can be grown by either LPE, VPE, or MBE. Sn doped GaAs films prepared by MBE exhibit a very low degree of compensation. From photoluminescence spectra taken at 2 K, there is no evidence that any incident Sn atoms are incorporated into acceptor sites. The measured free-donor concentration is thus directly proportional to the Sn flux in the molecular beams. High carrier concentrations up to 5×10^{19} cm^{-3} with excellent mobilities can be readily achieved. Assuming a residual acceptor concentration of 7×10^{15} cm^{-3}, electron mobilities at room temperature and 77 K as a function of free donor concentration agree rather well with the theoretical curve of Rode and Knight[137, 138)] (cf. Fig. 24). The photoluminescence inten-

Table 2. Promising n-type dopants in MBE GaAs

Element	Incorporation behaviour	Degree of compensation	Maximum achievable electron concentration	Remarks
Si	Predominantly donor	Fairly low	7×10^{18} cm^{-3}	Evaporation temperature fairly high (~1050 °C)
Ge	Depends strongly on growth conditions	Can be high	4×10^{18} cm^{-3}	Donor at high As_4 to Ga flux ratio and low growth temperature
Sn	Donor only	Low	5×10^{19} cm^{-3}	Tends to accumulate at the growth surface
S Se Te	Donors only	Low	8×10^{18} cm^{-3}	Molecular sources like PbS, PbSe, SnTe, PbTe or Tl_xTe_y for controllable incorporation required

sity for Sn-doped GaAs films grown by MBE was found to be greater than for high-quality Sn-doped LPE GaAs layers[139].

The only major disadvantage of using Sn is its tendency to accumulate at the GaAs surface during epitaxial growth. Using AES and SIMS, Ploog and Fischer[78] provided the first definite experimental evidence that the deviation from the ideal doping profile found in Sn doped MBE GaAs layers[60] is in fact caused by an enrichment of Sn in the outermost atomic layers during deposition. In all of the Sn-doped 1 μm thick GaAs layers grown by molecular beam epitaxy at substrate temperatures of 490 to 600 °C with carrier concentrations ranging from 5×10^{16} to 7×10^{18} cm^{-3} an unusually high amount of Sn in the outermost atomic layers of the samples was detected by AES, although growth conditions had been kept strictly constant during deposition. Fig. 20a shows the Auger spectrum of an as-grown Sn-doped GaAs film prepared by MBE at 550 °C. The Sn molecular beam was cut off 1 min before GaAs growth was stopped, in order to make sure that the last species arriving at the growth interface were Ga and As_4. As determined subsequently by Hall measurements and Raman scattering[65], the bulk material of the film had a *homogeneous* electron concentration of 6×10^{17} cm^{-3} and exhibited a mobility of 3200 cm^2/V · s at room temperature. This value of μ was in excellent agreement with the theoretical optimum[137, 138]. Therefore, it could be expected that for net carrier concentrations below 10^{18} cm^{-3} all Sn species incorporated into MBE GaAs were ionized, and that the total Sn concentration in the bulk film was not much higher than the measured carrier concentration.

For a uniform distribution, the Sn concentration of 5×10^{17} cm^{-3} corresponds to less than 10^{-5} monolayers. The detection limit of the particular AES measurements, however, was at best 10^{-2} monolayers. This implies that the peak found at 430 eV results from an accumulation of Sn in the outermost atomic layers of the growing GaAs film. The Sn enriched outermost atomic layers could be removed by a rather moderate Ar^+ ion sputtering. Figure 20b shows that no Sn was detectable after the removal of a 30 Å layer from the as-grown surface with 1 keV Ar^+ ions. Annealing the sputtered sample to 550 °C did not have any noticeable effect on the Sn distribution (Fig. 20c). This indicated that the surface enrichment with Sn in MBE GaAs is not caused by an after-growth diffusion of Sn to the surface. As a result of this behaviour, no abrupt doping profiles perpendicular to the growth surface can be achieved during MBE of GaAs, if Sn is used as donor impurity[60].

From careful experimental studies, Wood and Joyce[22] have developed a kinetic model to describe the incorporation behaviour of Sn. They observed that the Sn incorporation is surface rate limited. Before a steady-state free-donor concentration in the growing GaAs film can be achieved, it is necessary to build up a steady-state surface population of Sn which may be as large as 0.1 monolayers. In practise, this produces dips in the free-donor concentration profile at the film-substrate interface, which can be overcome only by predeposition of Sn prior to growth. According to Wood and Joyce[22], the Sn incorporation rate is simply determined by the concentration of Ga vacancies on the growth surface. The Ga vacancy concentration, on the other hand, should be directly proportional to the As_4 flux and inversely related to the substrate temperature. As a result, the incorporation-rate constant was found to be strongly dependent on the As_4 flux and to a lesser extent on the substrate temperature. Recently, Heckingbottom et al.[140] have proposed an alternative model to describe the dopant incorporation behavior, which is based on thermodynamic rather than on kinetic arguments. They derived a thermodynamic guideline for all common dopants in MBE GaAs including Sn, which fits surprisingly well with the experimental data.

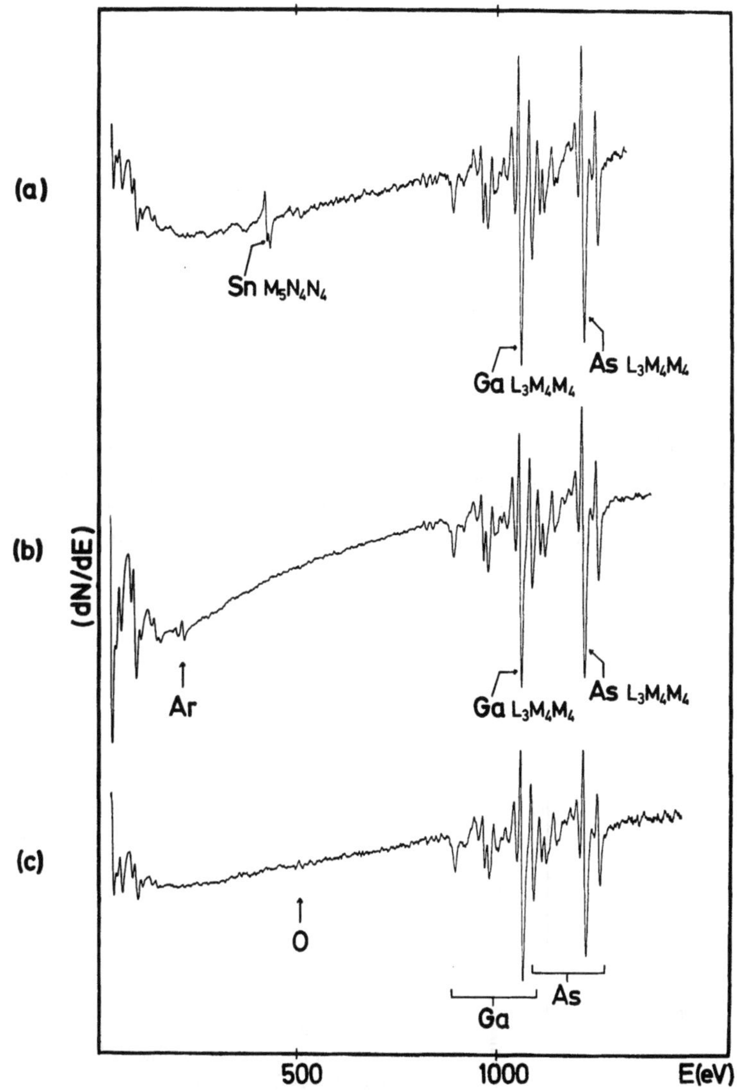

Fig. 20 a–c. Auger spectra of variously treated surface of a Sn-doped GaAs film grown by MBE ($n = 6 \times 10^{17}\ cm^{-3}$); (**a**) as-grown film indicating the surface segregation of Sn; (**b**) after removal of the outermost atomic layers by a moderate Ar^+ sputter etching; (**c**) after annealing the sputtered sample at 550 °C for 20 min

Sn forms shallow donors also in the MBE grown alloy systems $Al_xGa_{1-x}As$, $In_xGa_{1-x}As$[107], and $GaAs_ySb_{1-y}$ [107, 141] for $y \geqslant 0.2$. However, Sn acts as acceptor in $GaAs_ySb_{1-y}$ when $y \leqslant 0.2$[107].

Abrupt profiles and well resolved periodic profiles of donors in MBE GaAs can be obtained using Si or Ge as donor impurities[60] both of which are amphoteric. This means that the incorporation of the impurity element into either Ga or As sites depends critically on the As_4 to Ga flux ratio and on the substrate temperature. Under As-stabilized growth conditions and at substrate temperatures below 600 °C *germanium* is predominantly incorporated on Ga sites and acts as donor. Figure 21 shows measured and calculated values of free donor concentration as a funtion of Ge cell temperature obtained at constant growth rate (1.1 μm/h), substrate temperature (T_s = 500 °C) and a constant As_4 to Ga flux ratio of 2[142]. This Clausius-Clapeyron type plot demonstrates an excellent agreement over a wide range. The degree of auto-compensation does apparently not depend on the doping level. By analogy with LPE this incorporation behaviour can be described qualitatively by the following equation estimating the compensation ratio

$$\ln \frac{[Ge_{Ga}]}{[Ge_{As}]} = \ln \frac{X^l_{As}}{X^l_{Ga}} \frac{\Delta G^o_{A/B}}{R \cdot T} - (\ln\gamma^l_{Ga} - \ln\gamma^l_{As}) \qquad (4)$$

which was derived by Teramoto[143] for LPE growth of GaAs doped with either Si, Ge, or Sn. Here X^l_{As} and X^l_{Ga} represent the atomic fractions of arsenic and gallium in the melt and γ is the corresponding activity coefficient. In the case of Ge the value of $\Delta G^o_{A/B}$ was calculated to be −6.3 kcal/mol[143]. When Eq. (4) is applied to MBE growth, X is replaced by the partial pressures and $\ln\gamma_{Ga}$ and $\ln\gamma_{As}$ are omitted.

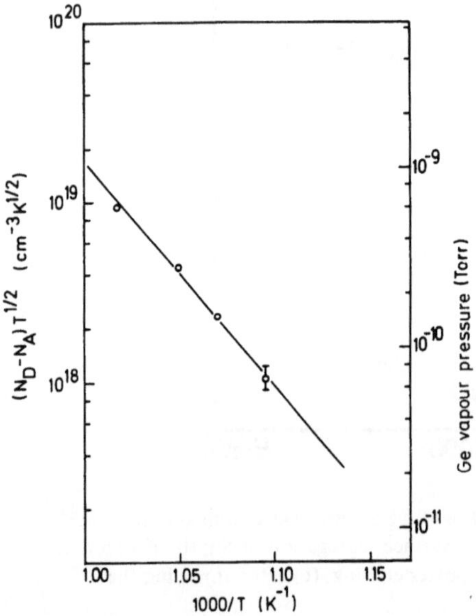

Fig. 21. Clausius-Clapeyron type plot of of free-donor concentration times $T^{1/2}$ in Ge-doped GaAs films as function of reciprocal Ge-effucion-cell temperature for constant film growth rate, substrate temperature and As_4 to Ga flux ratio. Also included is the Ge equilibrium vapor pressure curve

Table 3. Dependence of the site occupation of Ge in Ge-doped GaAs films on the growth temperature as reflected by the resulting carrier type, net carrier concentration, Hall mobility and specific resistance. The total amount of incorporated Ge atoms ($\sim 4.0 \times 10^{16}$ cm^{-3}) remained constant throughout all growth runs

T_S[°C]	n, $p_{300\,K}$[cm^{-3}]	$\mu_{300\,K}$[cm^{2}V^{-1}s^{-1}]	$\rho_{300\,K}$[Ohm cm]	Sample
500	$n = 3.3 \times 10^{16}$	4650	0.04	1029
575	$n = 2.8 \times 10^{16}$	4280	0.05	1032
595	$n = 1.4 \times 10^{16}$	2930	0.15	1033
610	$p = 4.8 \times 10^{15}$	319	4.07	1035
620	$p = 3.9 \times 10^{15}$	316	5.04	1034

Until recently, Ge was found[9, 99)] not to be particularly useful as a practical n-type dopant in MBE GaAs, because it had exhibited a high degree of auto-compensation when the films were grown under standard As-stabilized growth conditions. However, taking into account Eq. (4), Wood et al[61)] and Ploog et al.[100)] have now independently demonstrated that relatively uncompensated Ge-doped n-type GaAs films can be deposited with sufficiently large As_4 to Ga flux ratios, i.e. $J_{As_4}/J_{Ga} \geqslant 10$, and with low substrate temperatures, i.e., $T_s \leqslant 530$ °C. Under these conditions, the fraction of Ge atoms incorporated into Ga sites is greatly enhanced. In practice, it is highly desirable to perform MBE growth at flux ratios J_{As_4}/J_{Ga} between 2 and 1, in order to reduce the presence of non-stoichiometric centers in the grown films[144, 145)] associated with excess arsenic. Table 3 indicates that even under these growth conditions Ge-doped n-GaAs films with $N_D/N_A \approx 4/1$ exhibiting reasonable carrier mobilities can be grown at $T_s = 500$ °C. Also shown is the dependence of the autocompensation, which reflects the site incorporation on the substrate temperature, all other parameters remaining constant. Although the total amount of the incorporated Ge atoms remains constant, the 300 K free carrier concentration and mobility drastically decrease with increasing substrate temperature. At $T_s = 602$ °C roughly an equal amount of Ge is incorporated on Ga and As sites. Above this temperature, a larger fraction of Ge occupies As sites, thus yielding p-type material.

The increase of the As site occupation by Ge in n-type films with increasing substrate temperature is also reflected by the 2 K near band-gap photoluminescence spectra depicted in Fig. 22[142)]. The presence of background carbon acceptors in nominally undoped MBE GaAs films (cf. Sect. 3.3) with a concentration of about 1×10^{15} cm^{-3} is indicated by the strong peak at 1.493 eV which is assigned to a conduction-band-carbon acceptor (BA) transition. For clarity, the intensities of all peaks are normalised to the intensity of the BA transition due to carbon. It is obvious that the intensity of the peak at 1.479 eV corresponding to the BA emission involving Ge acceptors[175)] ($\sim$40 meV above the valence band) increases rapidly with increasing substrate temperature which also indicates an increasing auto-compensation.

The data presented in Table 3 indicate that Ge may become a very promising p- as well as n-type dopant for MBE GaAs. Since the procedure of changing the substrate temperature can be achieved within seconds and monitored in the RHEED

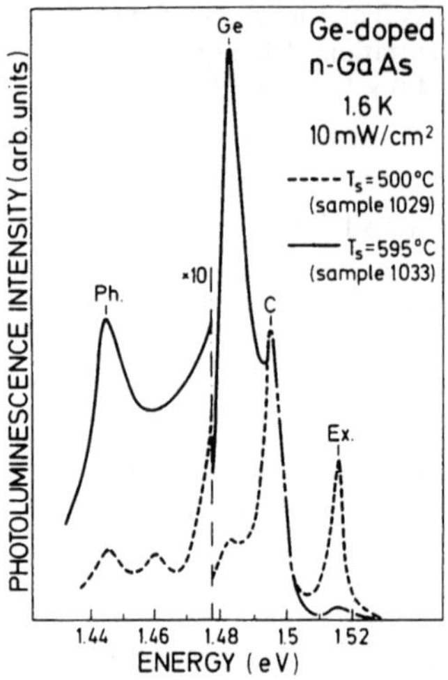

Fig. 22. Photoluminescence of two Ge-doped n-type GaAs layers arranged in the same order of increasing compensation ratio as summarized in Table 3. For clarity the intensity of all peaks are normalized to the peak height of the carbon BA transition

pattern, abrupt p/n junctions in GaAs with well defined and smooth interfaces can now be fabricated in a highly reproducible manner[100)].

Although *silicon* is an amphoteric dopant for LPE grown GaAs, in GaAs grown by MBE it is mainly incorporated on Ga sites yielding n-type material of fairly low compensation. Under As-stabilized growth conditions, Cho and Hayashi[146)] have readily obtained free donor concentrations from 1×10^{16} to 5×10^{18} cm^{-3} at substrate temperatures ranging from 450 to 580 °C. They used, however, a single polycrystalline GaAs source for the Ga as well as for the As_2 beam. For useful dopant fluxes $\geqslant 10^9$ atoms/cm^2 · s obtained by sublimation, the Si source must be heated to temperatures $\geqslant$1000 °C. Therefore, the material used for the Si effusion cell must be carefully selected, since cells operated at T $\geqslant$ 1000 °C are potential sources of unintentional impurities.

The final potential dopant of the group IV elements, *lead,* has a unity sticking coefficient on GaAs at and slightly above room temperature, but it desorbs readily at temperatures $\geqslant$250 °C. Cho[147)] and Wood et al.[136)] have recently found that Pb is not detectably incorporated in GaAs grown by MBE at substrate temperatures above 480 °C. In addition, GaAs films grown under high Pb fluxes in excess of five times that of the Ga flux show no degradation of the electrical and structural properties.

Although elemental *tellurium* has a sticking coefficient close to unity on GaAs, it was found[1, 71)] to be unsuitable as a dopant for MBE growth. Te tends to accumulate on the surface by displacing arsenic atoms from the As-stabilized growth surface and forming a stable surface compound, presumably GaTe. The high surface concentration of Te built up on GaAs during growth makes the doping process extremely difficult to control.

By means of a synergic reaction using SnTe, PbTe or Tl_xTe_y as molecular tellurium sources (even in the vapor phase), an efficient and controllable incorporation of Te as an n-type dopant into GaAs during MBE growth has been achieved[135)] yielding relatively uncompensated material with promising doping profiles up to

$n \geqslant 10^{18}$ cm^{-3}. Taking into account the interaction behavior of Pb with GaAs at substrate temperatures above 480 °C, Wood[136)] has recently utilized PbS and PbSe in a surface exchange reaction as so-called captive sources of S and Se atoms for efficient n-type doping of GaAs. From these preliminary results it can be expected that the use of such molecular dopant sources during MBE growth will yield variously doped compound semiconductor thin films exhibiting highly promising transport properties.

3 Evaluation and Properties of MBE Grown Layers

Molecular beam epitaxy has been developed within a few years from surface physics experiments of purely scientific interest to a well established epitaxial process of increasing importance to planar device fabrication technology. The performance of many high-frequency and light-emitting devices requires crucial growth control characteristics which are particularly provided by the MBE technique. These device applications require precise control of layer uniformity and thickness in the submicron range as well as close tolerances in composition and in carrier density of predetermined doping profiles. Any advances in the electronic and optical quality of MBE grown layers depend upon rapid and pertinent evaluation of the material.

The range of material properties with which the crystal grower as well as the device designer are mainly concerned include carrier profiles, compensation, deep level concentration, carrier lifetime and diffusion length. Of special interest are nondestructive methods particularly suitable for very thin ($\leqslant 5$ μm) layers, the growth of which is the real strength of the MBE technique, such as phase contrast and scanning electron microscopy for examination of surface morphology and layer thickness; Hall effect, diode capacitance, DLTS, and Raman measurements to obtain values of the concentration, mobility, compensation ratio, deep levels, and profiles versus depth of the majority carriers; and finally photoluminescence (PL) measurements (involving minority carriers) for detection and recognition of residual and intentional impurities in the layer. Ideally, for analysis of the epitaxial layer a technique should be selected using those properties basic to the operation of the device which is subsequently fabricated from that particular material.

3.1 Surface Morphology

Uniform GaAs and $Al_xGa_{1-x}As$ layers are now routinely grown by MBE on substrate areas as large as 30 x 30 mm. When grown with an As-stabilized (2 x 4) or C(2 x 8) surface reconstruction at substrate temperatures ranging from 450 to 610 °C, a smooth mirrorlike surface appearance similar to that shown in Fig. 23a and b is generally achieved. In order to ensure very uniform deposition over a large substrate area, the substrate must be positioned far from the sources ($\geqslant$ 100 mm) so that the angular variation in beam intensity ($\sim\cos\theta$) is minimal across the substrate. Any

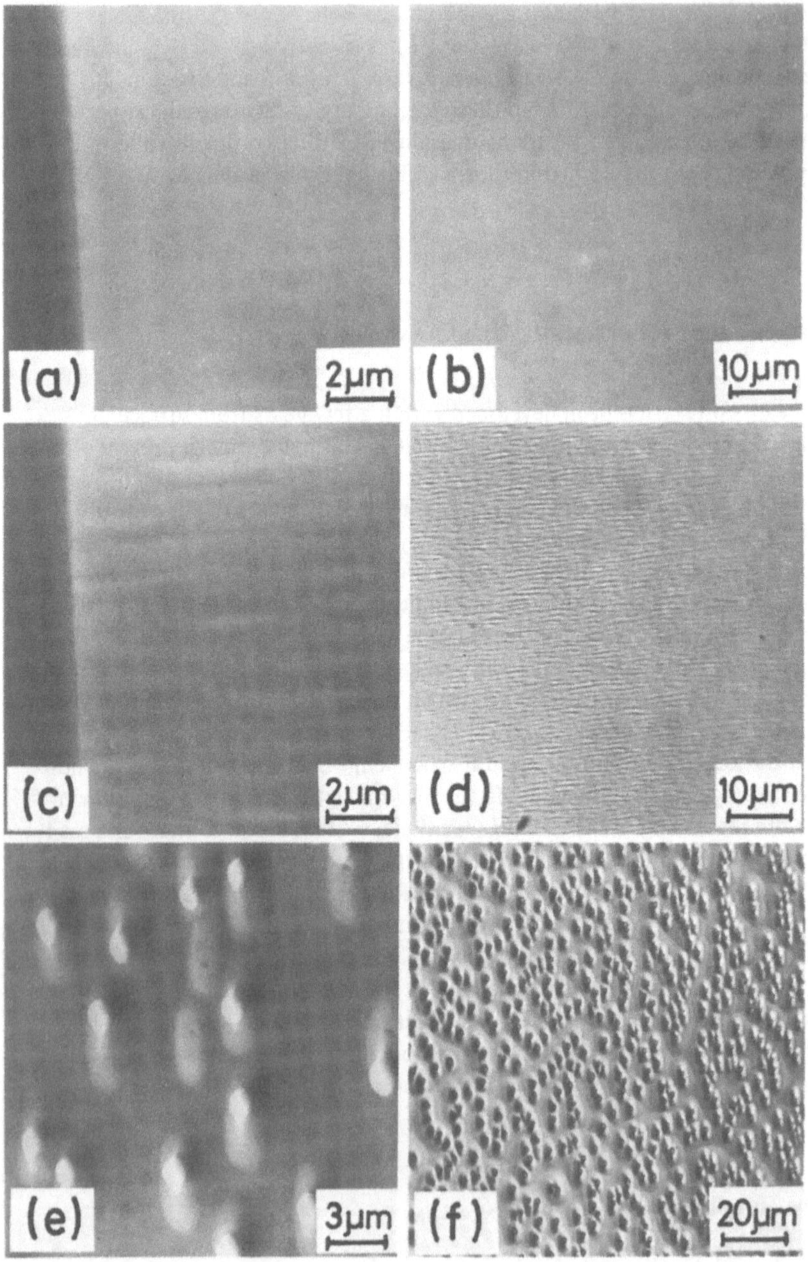

Fig. 23 a–f. Surface morphology of (100) GaAs grown by MBE under various conditions; (**a**) and (**b**) film grown with As-stabilized C(2 × 8) or (2 × 4) surface reconstruction; (**c**) and (**d**) film grown under the same conditions as wafer (**a**), however, on a substrate surface contaminated with ~0.1 monolayer carbon; (**e**) and (**f**) film grown with Ga-stabilized C(8 × 2) or (4 × 2) surface reconstruction indicating the buildup of Ga droplets. Figures **a**, **c**, and **e** are SEM micrographs with 45° specimen tilt. Figures **b**, **d**, and **f** are Nomarski interference contrast micrographs

increase in substrate-source distance drastically reduces the growth rate ($\sim d^{-2}$), which may, however, be advantageous from the point of view of precise control of layer thickness and doping uniformity. For areas of 30 x 30 mm less than 5% deviation from lateral uniformity across the substrate has been achieved.

Any distinct surface features detectable on MBE grown material usually result from residual substrate surface contamination which is shown in Fig. 23c and d for the case of carbon contamination on GaAs. They may also result from faulty epitaxial conditions including lattice mismatch between substrate and epitaxial film in the case of heteroepitaxy, growth temperature, deposition rate, and III–V flux ratio in the molecular beams. In the case of GaAs growth, the practical use of conditions with a Ga-stabilized (4 x 2) or C(8 x 2) surface reconstruction always yields samples with dull surfaces, in some instances with evidence of buildup of Ga droplets as shown in Fig. 23e and f.

3.2 Electrical Properties

Right from the beginning, MBE grown layers have been extensively characterized by the values of their majority carrier concentration and mobility obtained from Hall effect and conductivity measurements on uniformly doped layers. In practice, Hall constants R_H and resistivities ρ are measured by the van der Pauw[148)] method on symmetrically shaped samples with a magnetic field of 5 kG. In the author's laboratory cloverleaf patterns are only used for special purposes. Hall mobilities μ_H and carrier concentrations n can be calculated from $\mu_H = R_H/\rho$ and $n = r/e \cdot R_H$, where e is the electronic charge and the scattering factor r is assumed to be equal to 1. This assumption is an accepted compromise which rarely introduces an error greater than about 30%[149, 150)].

A temperature of 300 K is generally sufficient for nearly complete ionization of the relevant shallow donors and acceptors in GaAs. Thus, for shallow level impurities (say within ~50 meV of the band edge) the carrier density measured at room temperature by the Hall effect is equal to the net impurity density ($N_D - N_A$) or ($N_A - N_D$). If the compensation (N_A or N_D, respectively) is small, this value gives a reliable indication of the total impurity content of the sample. However, if this is not the case then the reduction in mobility below the value for uncompensated material of the same carrier density (derived by calculation or empirically) is used to estimate the degree of compensation. Since the influence of impurity scattering on the mobility is generally greater at 77 K than at room temperature, measurements at this fixed point often supplement room-temperature data.

Nominally undoped GaAs films grown from elemental arsenic and gallium sources are generally found to be p-type with a background hole concentration of $N_A - N_D \approx 1 \times 10^{15}$ cm^{-3} and hole mobilities μ ranging from 360 to 400 cm^2/V · s at room temperature. The main residual impurity is carbon which is consistent with the presence of a shallow acceptor with energy around 26 meV found in the 1.6 K PL spectra of all MBE grown GaAs samples (cf. Fig. 32). If careful attention is given to a further reduction of spurious impurities usually contained in an UHV system (H_2O, CO etc.) by additional cryopanels and by a sample exchange load-lock system, the background carrier concentration may be reduced to $N_A - N_D \leqslant 5 \times 10^{14}$ cm^{-3}.

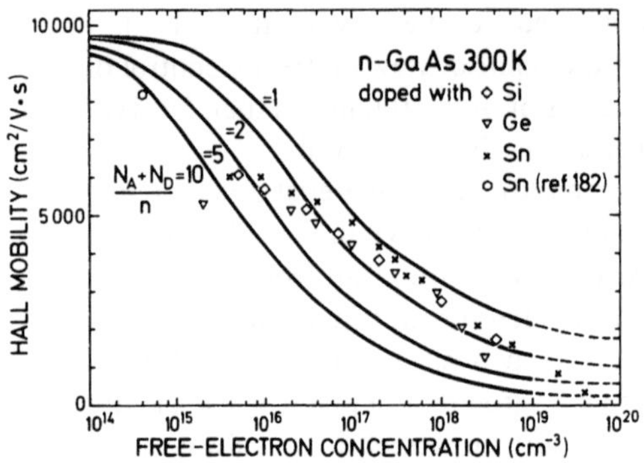

Fig. 24. 300 K Hall mobility as a function of free-electron concentration in MBE-grown GaAs films doped with either *Si, Ge,* or *Sn.* The solid lines follow the theoretical calculations of Rode and Knight[137)] for different compensation ratios

The source of carbon in MBE grown GaAs films is not quite clear. Carbon would be an ideal shallow acceptor in GaAs, if an intentional incorporation could be achieved. Therefore, Ilegems and Dingle[129)] deliberately introduced CO and C_2H_4 in the growth zone during GaAs deposition. Apparently, however, no change in the electrical and optical properties of the GaAs layers was achieved during these experiments. This was due to the low sticking coefficients of those neutral carbon species on GaAs. Fischer et al.[158)] recently obtained some results indicating that the Ga effusion cell itself may be the source of carbon contamination during MBE growth. If the materials of the cell are not carefully selected, a high-temperature reaction between the carbon crucible and the alumina support tube for the Ta heater wire (cf. Fig. 4) yields highly excited carbon containing species. These reactive species can well be incorporated into GaAs during growth, and carbon acceptor concentrations up to 8×10^{17} cm^{-3} have been readily achieved[158)]. Some of the electrical data of GaAs films in this way "intentionally" doped with carbon are included in Fig. 25 and 26.

Using GaAs as the arsenic source, the as-grown GaAs films always exhibit n-type conductivity in the mid 10^{15} cm^{-3} range with room temperature mobilities μ ranging from 5000 to 7000 cm^2/V · s. This residual conductivity is the result of Si contamination, since the commercially available undoped polycrystalline GaAs contains a large amount ($\sim 10^{16}$ cm^{-3}) of Si impurity.

In Fig. 24 the electron mobility μ at 300 K is plotted as a function of free-carrier concentration for various n-type GaAs films of 1–4 μm thickness doped with either Si, Ge, or Sn. The carrier concentration covers the range $5 \times 10^{15} \leqslant n \leqslant 5 \times 10^{19}$ cm^{-3}. For comparison, the theoretical variation calculated by Rode and Knight[137)] and by Rode[138)] for different levels of compensation is also plotted. This compensation ratio is defined as

$$\frac{\text{concentration of fixed ionized centers}}{\text{concentration of mobile charges}} = \frac{N_A + N_D}{n} .$$

For ideal extrinsic n-type GaAs, where the acceptor concentration N_A is negligible, the compensation ratio becomes unity. This case corresponds to the upper curve of Fig. 24. For a given electron concentration, Si- and Sn-doped MBE GaAs layers exhibit slightly higher mobilities than those doped with Ge. This effect is

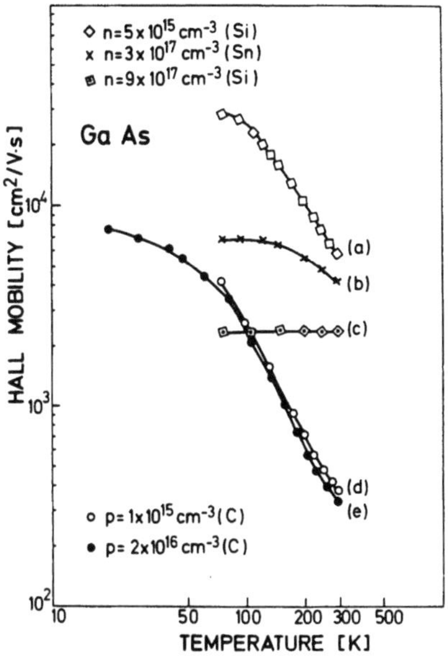

Fig. 25. Hall mobilities as function of temperature for some representative p- and n-type GaAs films grown by MBE

caused by the strong amphoteric character of Ge, which produces an increasing autocompensation at total impurity concentrations $\geqslant 8 \times 10^{17}$ cm^{-3}. For all three dopants, the deviation of experimental values from the theoretical curve indicating an increasing compensation ratio is significant at low ($n \leqslant 1 \times 10^{16}$ cm^{-3}) and of course at rather high carrier concentrations ($n \geqslant 1 \times 10^{18}$ cm^{-3}). When n is below 1×10^{16} cm^{-3}, the background carbon acceptor has a significant influence on the majority carrier mobility. Thus, the experimental data can be fitted rather well to a theoretical curve based on Rode's derivation when a constant acceptor concentration of $N_A \approx 7 \times 10^{15}$ cm^{-3} is assumed (not included in Fig. 24). In the limit of large free-electron concentrations ($n \geqslant 2 \times 10^{18}$ cm^{-3} for GaAs) the semiconductor degenerates. The neglect of degeneracy in the theoretical curves of Fig. 24 overestimates the mobility at room temperature by 25% for $n = 5 \times 10^{18}$ cm^{-3} and by 65% for $n = 1 \times 10^{19}$ cm^{-3} when $\frac{N_A + N_D}{n} = 1$.

Figure 25 shows the temperature dependence of the Hall mobility μ for several representative MBE-grown GaAs samples. Curves a, b and c were obtained from n-type samples and curves d and e from p-type samples. For n-GaAs a meaningful analysis of the temperature dependence of the Hall constant as well as of the Hall mobility has been established by Wolfe et al.[151)] and by Stillman and Wolfe[149)]. The authors analyzed the Hall mobility as a function of temperature with the Brooks-Herring[152)] equation for ionized impurity scattering. This impurity scattering mechanism dominates at liquid nitrogen temperature, and in samples with $N_D \geqslant 10^{16}$ cm^{-3} at even higher temperatures. Generally, the mobility in perfectly pure (intrinsic) GaAs is governed by three important types of scattering mechanism [(i) polar optical-mode scattering, (ii) acoustic-mode and piezoelectric scattering, and (iii) ionized-impurity scattering] which become dominating in different temperature ranges. The curve a in

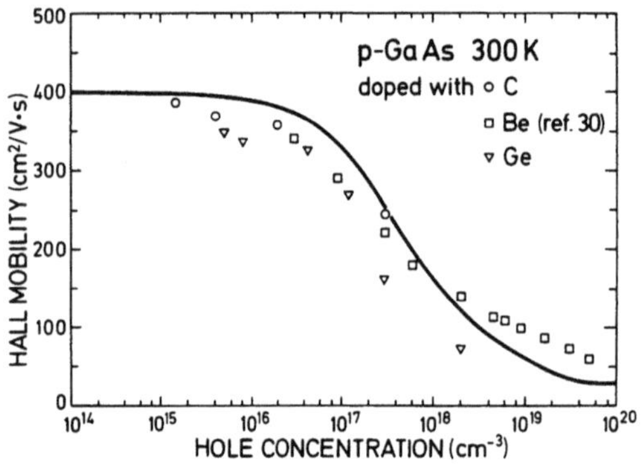

Fig. 26. 300 K Hall mobility as function of net-hole concentration in MBE-grown GaAs films doped with either *C, Be,* or *Ge.* The solid line represents the empirical curve proposed by Wiley[157)]

Fig. 25 obtained from an n-type layer illustrates a pronounced variation of mobility as a function of temperature indicating the dominating influence of different scattering mechanisms in different temperature regions. In samples with donor concentrations $N_D \geqslant 10^{16}\ cm^{-3}$ (curves b and c in Fig. 25) the impurity scattering mechanism dominates at even higher than liquid-nitrogen temperatures, and the mobility-vs.-temperature dependence becomes less distinct.

Significantly higher electron mobilities (up to a factor of two greater at room temperature) were recently obtained in MBE-grown $GaAs/Al_xGa_{1-x}As$ layer structures in which the mobile carriers and their parent donor impurities are spatially separated from each other[153–156)]. Particularly at low temperatures this charge separation leads to electron mobilities exceeding the conventional Brooks-Herring predictions. Moreover, these layer structures exhibit exciting two-dimensional properties which are more fully discussed in Sect. 4.2.

Fig. 26 shows the dependence of the room-temperature Hall mobility on the hole concentration for several p-type GaAs films of 1–4 μm thickness doped with Be and Ge. Also included are three values of nominally undoped as-grown p-type GaAs films with residual carbon acceptor. Because of the complexity of transport in the valence bands, a proper theoretical model for fitting Hall mobility data in p-type III–V compounds has not yet been established. The solid line shown in Fig. 26 was adopted from Wiley[157)] who simply calculated the dependence of mobility on hole concentration at 300 K using the Brooks-Herring formula together with an assumed lattice-limited mobility of 400 $cm^2/V \cdot s$. The author emphasized, however, that this procedure does not have any real theoretical significance. It simply provides a curve which follows the general trend of the experimental data obtained from the best p-GaAs samples prepared by VPE and LPE. The data points of the nominally undoped p-GaAs layers grown in the laboratory of the author and the Be-doped GaAs layers compiled by Ilegems[30)] agree reasonably well with the highest mobilities obtained from VPE and LPE samples. Again the strong amphoteric character of Ge is indicated by an increasing auto-compensation in p-type MBE GaAs at $p \geqslant 1 \times 10^{18}\ cm^{-3}$.

Post-growth heat treatment of the layers within the growth chamber in various atmospheres such as As_4, O_2, or N_2[158)], or the introduction of reactive gases during growth can drastically modify both the background level and the mobility in MBE grown material. The effect of introducing a partial pressure of hydrogen during growth on GaAs was recently reported by Calawa[159].

In many instances the average values of impurity concentration obtained from Hall measurements provide insufficient information for adequate material characterization, and particularly for thin layer structures the depth distribution must also be known. The net impurity concentration as a function of depth near the surface of a semiconductor, $n(X) = N_D(X) - N_A(X)$, can be derived from an analysis of the capacitance-voltage (C-V) relationship of a reverse biased Schottky barrier or an abrupt p/n junction. Hilibrand and Gold[160] have shown that the local net impurity concentration at X is given by

$$n(X) = \frac{2}{\epsilon \cdot e \cdot A^2} \left(\frac{d(1/C^2)}{dV}\right)^{-1} , \tag{5}$$

where e is the electronic charge, ϵ is the dielectric constant, C is the capacitance, A is the diode area, and V is the reverse bias. For depth profiling of non-uniform material, the result is more conveniently written in the following form

$$n(X) = -\frac{C^3}{\epsilon \cdot e \cdot A^2} \left(\frac{dC}{dV}\right)^{-1} . \tag{6}$$

Several experimental approaches to C-V profiling and extracting the carrier density as a function of depths have been described in the literature. The best known are the Copeland[161] profiler which is based on constant current probing and the Miller[162] profiler which uses a constant electrical field probing principle. Recently, Künzel et al.[163] introduced a versatile experimental setup for C-V profiling which is directly attached to the MBE equipment.

The formation of a satisfactory Schottky diode is essential to the successful use of the C-V technique. In the case of GaAs, evaporated gold dots generally exhibit excellent barrier characteristics. For a rapid on-line monitoring of as-grown GaAs samples, mercury probes to make the barriers have been successfully used[164]. Recently, Cho and Dernier[165] demonstrated that single crystal Al Schottky barrier diodes can be epitaxially grown on (100) MBE GaAs at room temperature in the same growth run. These diodes exhibit nearly ideal electrical characteristics with barrier heights of 0.66 eV and 0.72 eV for diodes formed on As-stabilized and Ga-stabilized surfaces, respectively, while for air-exposed GaAs surfaces the barrier height is of the order of 0.80 eV[166, 167]. A similar result has been reported for the differences in the barrier heights of Ag contacts on variously treated (100) GaAs and (100) InP surfaces[168, 169].

To minimize problems arising from a high series resistance, the C-V characteristics are usually determined in films grown on n^+ substrates. This precludes the possibility of carrying out Hall and conductivity measurements on the same sample. However, these measurements are made on a film grown concurrently on a semi-insulating substrate wafer which is mounted on the Mo heater block alongside the n^+ substrate.

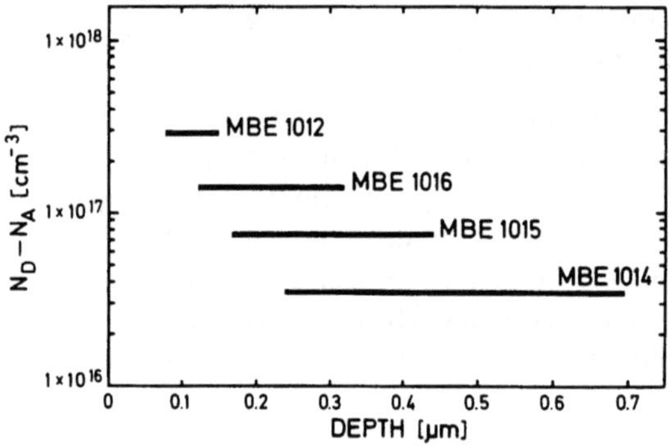

Fig. 27. Net donor concentration as function of depth for four representative Ge-doped GaAs films as derived from C-V measurements

The uniformity or the predetermined shape of doping profiles in epitaxially grown layers are crucial to the operation of most microwave devices. Figure 27 shows the net donor concentration as a function of distance into the film for four typical MBE-grown GaAs samples doped with Ge[142]. The free-donor concentration profiles are resonably flat over the measurable part of the film. The outermost region of the film cannot be measured because of the zero-bias depletion layer arising from the Schottky-barrier height (~0.98 eV for Au[166]). The depletion width varies from ~0.3 µm for $n = 2 \times 10^{16}$ cm^{-3} to ~0.03 µm at $n = 2 \times 10^{18}$ cm^{-3}. Furthermore, the region of the film closer to the substrate interface cannot be measured, because the maximum charge which can be depleted before breakdown occurs, due to the high electric field at the surface, is limited to about 2×10^{12} carriers cm^{-2}. Thus, the technique of diode capacitance profiling is confined to a (depth x doping density) product of $\sim 2 \times 10^{12}$ cm^{-2}. However, by successive anodic stripping using an aqueous $(NH_4)H_2PO_4$ solution followed by removal in HCl, Wood et al.[22, 61] have profiled a wide range of doping concentrations over an unlimited depth in MBE GaAs layers. The capability of diode capacitance profiling is demonstrated by means of Fig. 19 which illustrates measured carrier concentration profiles in a reverse-biased n^+–p–p^+ junction with periodic dopant modulation in the p region. The resolution of such doping profiles is on the order of two or three times the Debye length. In the case of GaAs, for carrier concentrations of 10^{16} and 10^{17} cm^{-3}, the Debye lengths are ~430 and ~136 Å, respectively.

The reverse breakdown voltage of GaAs p/n junctions yields important material parameters in relation to the performance of many semiconductor devices which require such structures. Recently, improved p/n junctions in Ge-doped GaAs consisting of p-type layers with extremely smooth surfaces and low-compensation n-type layers were fabricated[100]. During growth the site occupancy of Ge was controlled by varying the substrate temperature from $T_S \approx 500$ °C yielding the n-type layer, to $T_S \geqslant 600$ °C yielding the p-type layer at a *constant* As_4- to Ga-flux ratio of two. Thus, the change in doping between layers was controlled by the substrate temperature and monitored in the RHEED pattern. Representative current-voltage (I–V) characteristics of mesa diodes fabricated on such a structure are shown in Fig. 28. The area of the diodes was about 2×10^{-3} cm^2. In the forward direction for voltages above 0.4 V, the I–V characteristic of this n^+/p diode can be fitted by the equation

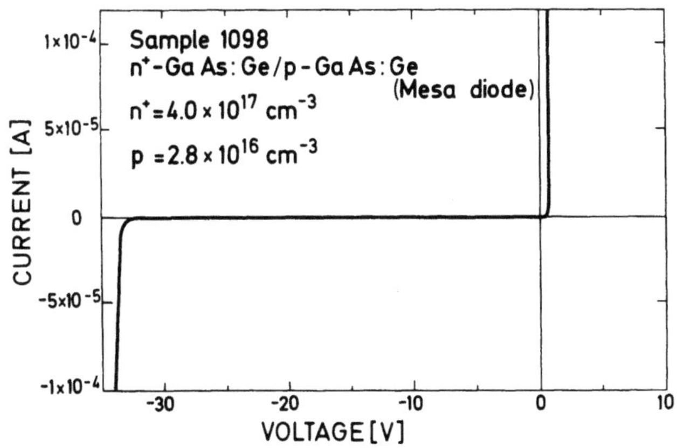

(a)

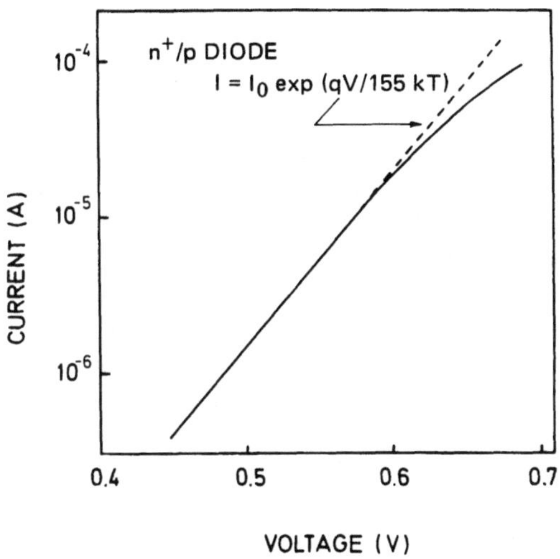

(b)

Fig. 28 a and b. Junction behavior of a representative Ge-doped GaAs diode at 300 K; **(a)** I–V characteristic; **(b)** exponential I–V dependence of the n^+/p diode in the forward direction

$I = I_S \exp(q \cdot V/n \cdot kT)$ with $n = 1.55$ (Fig. 28b), where I_S is the saturation current, q is the electronic charge, n is the diode non-ideality factor, and k is the Boltzmann constant. This value of n is expected if diffusion and recombination mechanisms are both comparable to the current across the p/n junction[166)].

The above mentioned techniques for transport measurements require physical contacts to the semiconductor samples. Strictly speaking, this is destructive since it produces at least some surface damage. The only method for transport measurements which does not need a direct electrical contact to the specimen is Raman scattering. Lattice vibrations provide a probe of crystal structure, of defects and of impurities with a scale of 2 to 4 Å. Using this technique, Abstreiter et al.[65)] have recently demonstrated that crystal orientation, carrier concentration, scattering times of

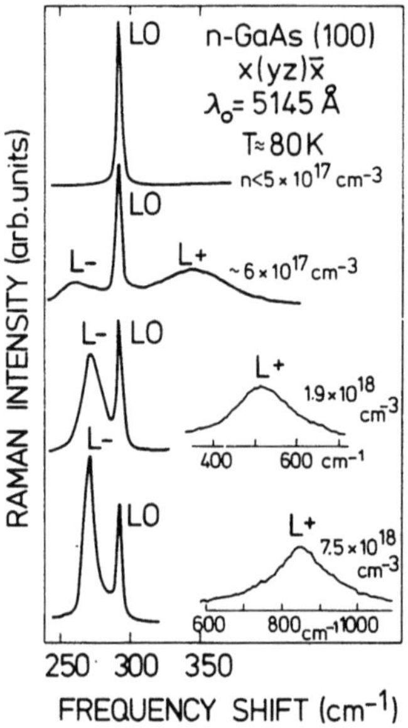

Fig. 29. Raman spectra of n-type MBE GaAs films with different donor concentrations obtained at 80 K in backscattering geometry. Both the *LO* phonon mode from the depletion layer and the coupled *LO* phonon-plasmon modes L_+ and L_- are observed

charge carriers, composition of alloy crystals and depth profiles can be studied in thin layers and heterostructures of GaAs and $Al_xGa_{1-x}As$. In Fig. 29 typical Raman spectra of n-doped GaAs layers grown by MBE are shown for four different carrier concentrations. The spectra were obtained at 80 K in the standard backscattering geometry on as-grown (100) surfaces with the emission line of an Ar^+ laser. In GaAs samples with $n \geqslant 5 \times 10^{17}$ cm^{-3} the coupled LO phonon-plasmon modes from the bulk of the layer and the unscreened LO phonons from the surface depletion layers can be observed simultaneously. The intensity of the bulk signal (L_+ and L_-) increases compared to the surface signal (LO phonon) with increasing carrier concentration. This is caused by the decreasing depletion width, while the penetration depth of the incident laser light remains unchanged (~0.1 μm at $\lambda_0 = 5145$ Å). For $\lambda_0 = 5145$ Å the observability of the L_+ and L_- signal is thus limited to carrier densities $n \geqslant 5 \times 10^{17}$ cm^{-3}. In this range the frequency of the L_- mode depends only very weakly on the carrier concentration, while the L_+ mode shifts quite strongly to higher energies when n is increased. In samples of unkonwn carrier concentration, the value of n can be determined from the position of this high-frequency L_+ mode using the relation[170)]

$$\omega_{L_+} = \left(\frac{4\,\pi n e^2}{m^* \cdot \epsilon_\infty}\right)^{1/2} , \qquad (7)$$

where ω_{L_+} is the frequency of the L_+ mode, e is the electron charge, m* is the effective mass of the carrier, and ϵ_∞ is the high-frequency dielectric constant. Excellent agreement is obtained with results of Hall and C-V measurements. When infrared frequencies are used for excitation, the carrier range can be extended down to $n \approx 1 \times 10^{16}$ cm^{-3} by studying the frequency of the L_- mode, which depends strongly on n in the range $1 \times 10^{16} \leqslant n \leqslant 5 \times 10^{17}$ cm^{-3}. In this case the depletion layer is no problem, since GaAs is transparent in the infrared region.

At high carrier concentrations ($n \geqslant 10^{18}$ cm^{-3}) the L_+ mode (Fig. 29) is plasmon-like, and the scattering time τ of the charge carriers can be estimated from the line width of this mode. Using the relation $\mu = e\tau/m^*$, the mobility of the charge carriers can also be estimated from the width of the L_+ mode by comparing the results with values of known standard samples. An inhomogeneous donor distribution would involve a highly asymmetric L_+ mode. A variation of the donor concentration with distance from the surface can be evaluated by using different laser excitation lines covering the range 1.83–2.81 eV, in which the penetration depth into GaAs changes from 0.3 to 0.03 μm.

Besides the fact that no contacts are required, the main advantage of Raman scattering compared to conventional characterization methods is that the carrier concentration can be determined within very small areas of the sample. The incident laser beam can be focussed down to 4 μm^2. For single GaAs wafers routinely grown in the laboratory of the author, the values of n were found to be within 3% of the average value indicating excellent lateral uniformity across the entire 20 x 15 mm area of the substrate.

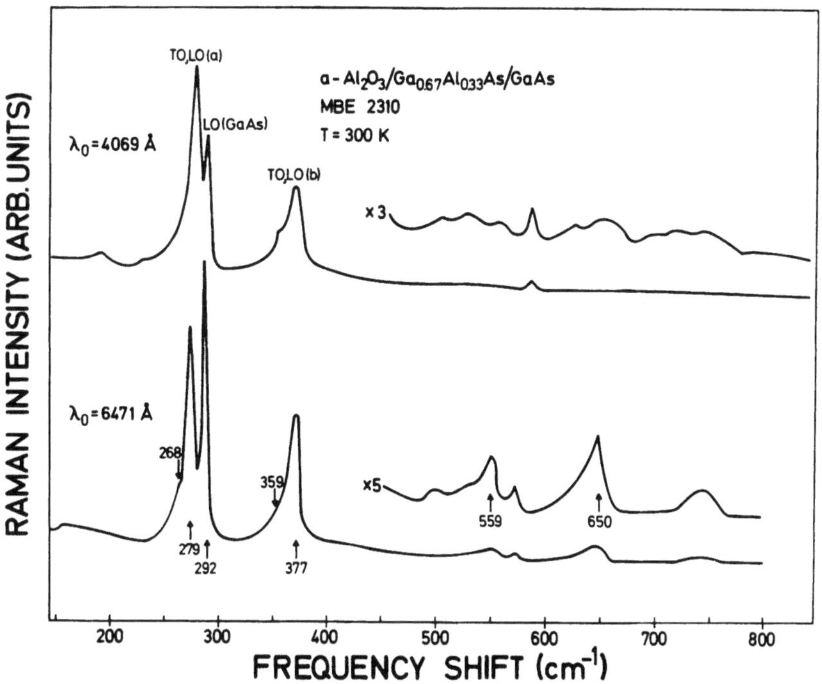

Fig. 30. Raman spectra of an a-$Al_2O_3/Al_{0.33}Ga_{0.67}As$/n-GaAs heterostructure obtained at 300 K in backscattering geometry with two different laser excitation lines. The $Al_xGa_{1-x}As$ and the GaAs layer can be clearly separated from the varying depth of information. The peaks above 400 cm^{-1} arise from second-order scattering

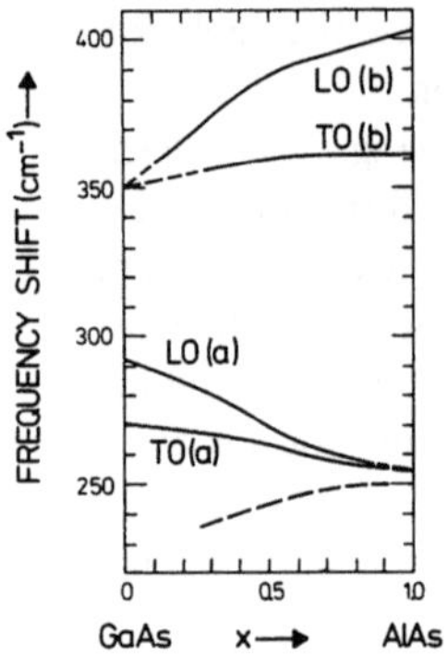

Fig. 31. Frequencies of the vibrational modes in $Al_xGa_{1-x}As$ as function of the molar fraction x (from Ref. [171])

In the $Al_xGa_{1-x}As$ alloy system the Raman spectra obtained in backscattering geometry from as-grown (100) surfaces show essentially two distinct bands in the whole composition range (Fig. 30). One appears at frequencies close to the modes of pure GaAs and the other close to pure AlAs. In Fig. 31 the dependence of the vibrational modes on the molar fraction x of Al in $Al_xGa_{1-x}As$ is depicted[171]. The frequency variation of these modes can thus be used to determine the value of x in unknown samples. With Raman scattering an accuracy of about 2% can be achieved for the determination of the Al concentration in the range $0.05 \leqslant x \leqslant 0.90$. In $GaAs/Al_xGa_{1-x}As$ multilayer systems without any special periodicity so that no zone folding occurs, Raman scattering yields the vibrational spectra of each individual layer, when different laser excitation lines with varying penetration depths are applied. For an $a\text{-}Al_2O_3/Al_{0.33}Ga_{0.67}As/GaAs$ heterostructure this effect is shown in Fig. 30[25]. This method of determining the composition of III-V alloys by Raman scattering has recently been extended to other systems including $Ga_xIn_{1-x}P$, $Ga_xIn_{1-x}As$[172], and $Ga_xIn_{1-x}P_yAs_{1-y}$[173].

Alternative nondestructive methods for the determination of the composition in III-V alloys are double-crystal X-ray diffraction and Rutherford backscattering. For $Al_xGa_{1-x}As$ the value of x below the direct-indirect crossover has also been obtained by equating the energy of the maximum photoluminescence emission at room temperature to the alloy band gap E_g and calculating x from the empirical relation[81]

$$E_g(eV) = 1.424 + 1.247 \cdot x$$

3.3 Luminescence Properties

An important problem in developing a new crystal growth technique like MBE is that of identifying relatively low levels of residual impurities in the grown material. It is generally desirable to reduce such background levels well below those deliberately incorporated for device fabrication. The possibility of using high-resolution photoluminescence (PL) at liquid-helium temperature to identify several different donors and acceptors in a single measurement has been well established for the most relevant

III–V compounds[174–176]. Hole-electron pairs in the valence and conduction band are generated by photon excitation with an energy higher than the bandgap of the semiconductor. It is the process of radiative and nonradiative recombination of these carriers that characterizes the semiconductor material; the nature of the luminescence is determined by the donors and acceptors and the efficiency is limited by non-radiative recombination centers.

Thus, the PL response of MBE grown layers is useful for characterization of the quality of the material in addition to the electrical data. Aside from some early studies by Cho et al.[1, 56, 99, 117, 139, 146], however, only very few examples of PL measurements made on MBE samples have been described in literature. Ilegems et al.[29, 30, 129] performed a thorough PL study on acceptor incorporation in MBE GaAs. Recently, Scott and Roberts[177] presented some low-temperature PL data of MBE GaAs doped with either Be, Ge, or Sn obtained with high excitation powers. Also included were data of $Ga_xIn_{1-x}P$. As yet, however, no detailed PL spectra of MBE grown $Al_xGa_{1-x}As$ have been published, although this material plays an important role in optoelectronic device fabrication. Because of the reactive nature of the $Al_xGa_{1-x}As$ surface, it has turned out to be extremely difficult to grow $Al_xGa_{1-x}As$ of good optical quality by MBE. Oxygen-containing species in the growth chamber, such as H_2O and CO, have sticking coefficients on $Al_xGa_{1-x}As$ at least one order of magnitude higher than on GaAs. The adsorption of these species during growth has been correlated with the inferior PL efficiency. A reduction in partial pressure of these species by using a sample interlock system as well as additional cryopanels surrounding the substrate region has led to improved $Al_xGa_{1-x}As$ optoelectronic device structures (see Sect. 4.1.2) thus indicating a partial enhancement of the PL intensity.

Figure 32 shows low-temperature PL spectra of unintentionally doped MBE GaAs films grown in two different growth systems under similar conditions except for some minor differences in the As_4-to-Ga flux ratios[142]. Standard excitation con-

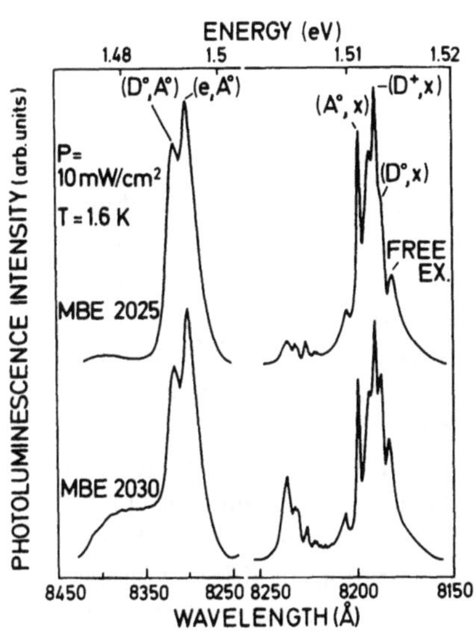

Fig. 32. Photoluminescence of two unintentionally doped GaAs films grown by MBE in different growth systems under As-stabilized conditions

ditions (Kr^+ laser with λ_0 = 6764 Å or 7525 Å at powers low enough not to disturb the distribution of conduction band electrons, i.e., $P < 30$ mW cm^{-2}) were maintained during all PL measurements so as to allow meaningful comparison between different crystals. The high energy features in the PL spectra of Fig. 32 are due to free exciton (~1.515 eV), donor bound-exciton [(D^+, x) at ~1.5135 eV], and neutral-acceptor bound-exciton ([A^0, x) at ~1.5120 eV] transitions. The double peaks at ~1.493 eV and ~1.491 eV are attributed to free-electron carbon-acceptor (BA) and donor carbon-acceptor (DA) pair transitions, respectively[178]. This doublet is not always resolved. Phonon replicas are also observed. In none of the PL spectra taken from unintentionally doped GaAs samples grown under As-stabilized conditions in the laboratory of the author a peak around 1.406 eV previously attributed to Mn on Ga sites[29, 129] has been observed.

There are several characteristic differences between the PL spectra of GaAs crystals grown by MBE and LPE or VPE. The most striking ones are the features in the energy range from 1.511 to 1.505 eV which can be resolved into seven peaks. This series of peaks has not yet been identified. However, they do occur in all PL spectra so far obtained from GaAs crystals grown from elemental arsenic and gallium, and they were also obtained from a MBE GaAs film grown at the Varian Corporate Solid State Laboratory[179]. Another result of the PL measurements is that the PL efficiencies of nominally undoped GaAs layers grown by MBE with the usual As-stabilized conditions are lower than those of LPE grown layers. It was shown that increased PL efficiencies can be obtained when the growth conditions are changed to Ga-stabilized, but the practical use of Ga-stabilized growth conditions is rather limited because of the surface roughening (see Fig. 23e and f). There has been speculation[29, 129] about the improvement in PL efficiency under these growth conditions, however, a satisfactory interpretation has not been reported. Room-temperature PL studies[139] of GaAs layers grown with the donors Si, Ge, and Sn have revealed that the luminescent intensity for MBE n-type layers depends on the growth temperature. The PL intensity for Sn-doped MBE GaAs layers was found to be greater than for high-quality Sn-doped LPE layers.

3.4 Deep Levels

Excess leakage current in reverse-biased p/n junctions or non-radiative recombination centers in light-emitting diodes and lasers are generally attributed to deep impurity levels acting as carrier traps or generation-recombination centers. The fact that deep impurities can produce such severe effects makes a control of their occurrence and a proper understanding of their properties highly desirable. Lang et al.[180] performed the only systematic study of electron traps in various n-type GaAs samples grown by MBE in six different growth systems with three different dopants (Si, Ge, and Sn) under both As- and Ga-stabilized conditions. The deep-level spectra were recorded by deep-level transient capacitance spectroscopy (DLTS). At least nine different electron traps were observed with energies ranging in depth from 0.08 to 0.85 eV from the conduction band. Total electron trap concentrations in different samples were found to vary from a maximum of 1.6×10^{14} cm^{-3} to a minimum in the mid

10^{12} cm^{-3} range. Studies of samples from various growth systems led the authors to the conclusion that most of the observed electron traps in MBE GaAs were caused by chemical contaminants and were not due to native defects induced by the special growth technique. Indeed, a thirty-fold reduction in trap concentration was observed from growth systems in which more careful attention was given to the reduction of spurious chemical contamination (H_2O, CO) by additional cryopanels around the substrate[180]. Finally, a comparison with high-quality epitaxial n-GaAs grown by LPE and VPE showed that all three types of growth techniques yielded distinctly different electron trap spectra. Three of the major traps reported by Lang et al.[180] corresponding to 0.19 eV, 0.30 eV, and 0.48 eV were also observed in MBE GaAs grown at Philips Research Laboratories[61].

For a reliable correlation between observed electrical properties and important features of crystal growth conditions it is not sufficient to identify each center electrically and measure its position in the energy gap, electron and hole capture cross sections, etc. The second important step would be to relate these observations to the introduction of specific impurity atoms or defect centers in order to obtain a more detailed understanding of either the physics or chemistry involved. As yet, however, only the first steps in this laborious direction have been undertaken for epitiaxially grown III–V compounds by combining deep-level measurements with, for example, electron-spin-resonance (ESR) measurements[181].

Nearly all studies on the properties of MBE grown material have been focussed on GaAs, because of its importance for device application. The MBE technique has now reached a state where, for GaAs, background levels in the low 10^{15} cm^{-3} range can be routinely achieved and with great care may be depressed to the 10^{14} cm^{-3} range. The highest room temperature mobility reported[182] for MBE grown n-type GaAs doped with tin is $\mu = 8084$ $cm^2/V \cdot s$ at a carrier concentration of $n = 4 \times 10^{14}$ cm^{-3}. The freeze-out ratio of the carrier in these samples as the temperature decreased to 77 K was reported to be only 1%. The lower limit of the total impurity concentration depends on source and system contaminants as well as on residual gas levels. This limit may not have been reached and further reduction should be possible in future experiments.

From all other binary and ternary III–V compounds which have so far been grown by MBE, few data on characteristic electrical properties are available. For unintentionally doped InP layers residual carrier concentrations ranging from $n = 2 \times 10^{16}$ to 2×10^{18} cm^{-3} have been reported[68], MBE grown InAs films were found to be also n-type with residual carrier concentrations in the 5×10^{16} to 1×10^{18} cm^{-3} range[106, 107, 183]. Similar values were obtained for the ternary $Ga_xIn_{1-x}As$ alloy[68, 69, 107], whereas nominally undoped films of GaSb and $GaAs_ySb_{1-y}$ ($0 \leqslant y \leqslant 0.7$) were found to be p-type[107]. $Ga_xIn_{1-x}P$ lattice matched to GaAs ($x = 0.53$) apparently contains deep donors in a concentration ranging from 8×10^{17} to 1×10^{18} cm^{-3} depending on the growth conditions[177].

4 Applications of MBE Grown Films

4.1 Device Applications of MBE

The performance of devices fabricated from the as-grown material can ideally be used as the best monitor for the progress of a new film growth technique. From the early beginning, microwave and optoelectronic devices have been fabricated from MBE-grown GaAs and $Al_xGa_{1-x}As$. During the past years this work has been concentrated mainly at Bell Telephone Laboratories. Fabrication and performance of various devices obtained from MBE material have been summarized in excellent reviews by Cho and Arthur[1] and by Cho[7, 184].

The unique feature of molecular beam epitaxy for device fabrication is the precise control of layer thickness, composition of ternary III–III–V alloys, and doping level for concentrations $\geqslant 5 \times 10^{15}$ cm^{-3}. Using mechanical shutters, molecular beams can be stopped and started with a few tens of angstrom of spatial resolution normal to the growth surface. These favorable characteristics have been exploited successfully in the fabrication of GaAs microwave devices including varactor diodes, low-high-low IMPATT diodes, mixer diodes, and Schottky barrier FET's, as well as in the fabrication of optoelectronic devices such as optical waveguides, light emitting diodes and heterostructure injection lasers.

4.1.1 Microwave Devices

For *varactor diodes* a tailored doping profile is required in order that small changes in reverse bias can produce large changes in capacitance as the depletion region widens. Cho and Reinhart[49] fabricated hyperabrupt varactors from n-type MBE GaAs doped with Sn, in which a bias change of less than 1 V produces a tenfold change in capacitance. Wood et al.[61] have recently demonstrated that also Ge-doped n-type MBE GaAs films can be successfully used in varactors for parametric amplifiers.

In low-high-low IMPATT (impact avalanche transit time) *diodes* a very abrupt change in doping level is required. The device is designed to convert dc to rf power with high efficiency. Cho et al.[185] produced GaAs films by MBE having a 2000 Å narrow spike of more heavily doped n-type region parallel to the growth surface. The IMPATT diodes fabricated from this material operated at a frequency of 11.7 GHz. With an output of 3 W, the diodes had an efficiency of 18% and a noise figure of 44 dB.

As the name implies, *mixer diodes* effect up or down frequency conversion by mixing the information-bearing signal with the signal of a local oscillator. Their basic structure consists of a Schottky barrier formed on an n-type active GaAs layer with a carrier concentration in the 10^{16} to 10^{17} cm^{-3} range and a thickness of 1000 to 3000 Å. The active layer is grown on an n^+ buffer layer ($n \approx 2 \times 10^{18}$ cm^{-3}) upon which the Ohmic contact is formed. Cho et al.[50, 186] developed a new method to form planar, electrically isolated GaAs devices called "MBE planar technology". Single crystal areas less than 5 μm on a side have been subsequently grown with this technique. The approach makes feasible the fabrication of isolated planar beam-leaded mixer diodes having parasitic capacitances less than half that of comparable devices fabricated by the standard mesa technique on semi-insulating substrates. When tested in a double-balanced down converter configuration[187], these mixer diodes exhibited less than 5.5 dB of conversion loss at their optimum frequency of 51.5 GHz. At 103 GHz their configuration loss was 8.5 dB.

Recently, Schneider et al.[188] fabricated millimeter wave mixer diodes from MBE grown GaAs with doping profiles specifically designed for low noise at cryogenic temperatures. The diode noise temperature of 55 K obtained under pumped conditions represents the lowest value obtained in resistive mixer. Using these diodes, a cryogenic millimeter wave resistive mixer has been built[189] which operated from 62 to 92 GHz with a system noise temperature below 480 K. This receiver was used in the 7-m offset Cassegrainian antenna at Bell Laboratories, Crawford Hill (USA), for studies of spectral-line radiation from molecules in the interstellar medium. A broad minimum in noise temperature was observed near 81 GHz where the system temperature was 312 K, the mixer temperature was 209 K, and the conversion loss was 6.7 dB (~5.9 dB after correction). The superior performance of these mixers was achieved by combining the MBE growth technique with a number of novel processing steps[188]. The devices are ideally suited for an application in cooled millimeter- and submillimeter-wave receivers.

The fabrication of both low-noise and high-power *GaAs Schottky barrier field effect transistors* (MESFET's) has been the most active area of research and development in the field of microwave devices for the last years. The reasons for all this activity are, first, that the GaAs FET exhibits higher gain and lower noise figure as a low-noise three-terminal amplifier over the frequency range 4 to 20 GHz than any other device. Secondly, the GaAs FET is similar in power but with higher gain as a power device when compared to Si bipolars at frequencies of 4 to 6 GHz and exceeds Si bipolars at higher frequencies.

The basic structure of a GaAs FET consists of a thin n-type film ($n = 5 \times 10^{16}$ to $5 \times 10^{17}\ cm^{-3}$, 0.3–0.7 μm thick) grown on a semi-insulating substrate (Fig. 33). Two crucial parameters are relevant to the active layer; first, the uniformity of the thickness and doping across the wafer and, secondly, the carrier mobility as a function of depth of the active layer. The uniformity can be estimated from the variation of the voltage V_K, which is required to pinch-off a Schottky diode on the material, across the wafer using the relation $V_K + 0.7 \approx 7 \cdot n \cdot x^2$ [cm], where n is the doping

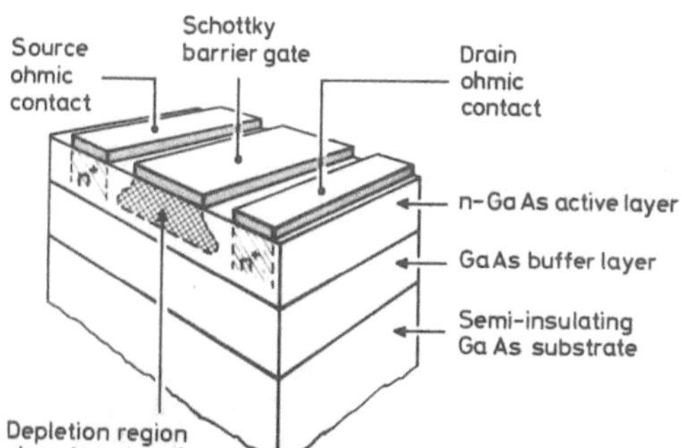

Fig. 33. Schematic illustration of the basic structure of a GaAs Schottky barrier field-effect transistor (MESFET)

level in cm^{-3} and x is the thickness in μm. A very uniform deposition over large substrate areas can be achieved with MBE when the substrate-to-source distance is increased to $d \geqslant 100$ mm. The dc characteristics (J_{DS}-V_{DS}) of GaAs FET's often display hysteresis loops, which are generally attributed to the alternate filling and emptying of traps at the substrate-epitaxial layer interface. This effect drastically deteriorates the carrier mobility.

Cho et al.[67] studied the variation of Hall mobility as a function of layer thickness when the layers were grown on chemically etched, ion-sputter-cleaned, and intermediate buffer-layer substrates. The authors found that the mobility of films directly deposited on the substrate drops at the interface independently of cleaning and increases slightly with buffer layers. Although reasonable results without buffer layers have once been reported[136], in general there is apparently a need even with the lower growth temperatures of MBE to include a p-type buffer layer between the active layer and the substrate for reproducible optimized device material. This produces devices, the I_{DS}-V_{DS} characteristics of which are essentially free of kinks or loops. The required doping profile of standard epitaxial layers used for GaAs MESFET fabrication is schematically shown in Fig. 34. The final n^+ layer is used for improved ohmic contact formation (see Sect. 4.1.3). The depicted three-layer structure is grown without interruption by appropriately adjusting the temperature of the dopant source (either Sn or Ge).

A new approach to improve the active layer-substrate interface has recently been presented by Morkoc and Cho[182] who first deposited a Cr- and Sn-doped semi-insulating GaAs layer followed by the Sn-doped n-type active layer. Improved linear dc characteristics over wide variations in gate voltage can be achieved by exponentially increasing the carrier profile of the active GaAs layer towards the interface to the buffer layer[190].

Both low-noise and high-power FET's have been fabricated on MBE GaAs layers[67]. For low-noise FET's, a noise figure of 1.9 dB with a corresponding gain of 11 dB was achieved at 6 GHz. The transductance was 40 mmho for devices with 500 μm gate-length. The power FET's with 3 mm gate-width had an output of 1.3 W (1 dB compression) at 4.4 GHz with a corresponding small-signal gain of 10 dB. The power added efficiency was 35%. Similar characteristics for power FET's were re-

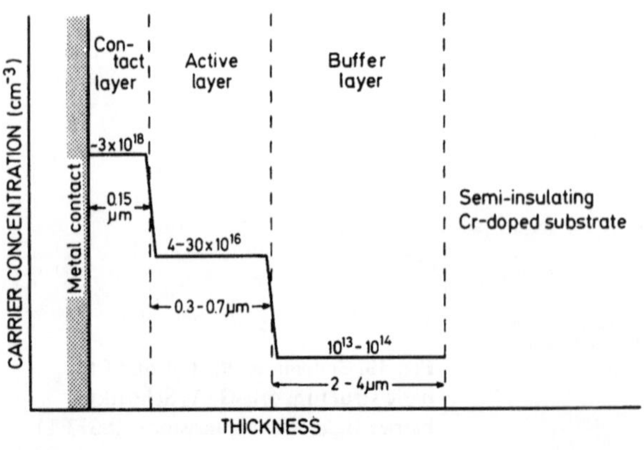

Fig. 34. Schematic doping profile of epitaxial GaAs films used for MESFET's

cently reported by Wataze et al.[250]. These results compare reasonably well with FET's obtained from VPE material[191, 192]. Thus, MBE shows considerable promise for the production of very high frequency GaAs MESFET's with accurately tailored parameters (shorter gate lengths, and thinner channels). A combination with selected area growth[186] may have strong impact on integrated FET's.

4.1.2 Optoelectronic Devices

For applications in optical devices there is great current interest in the materials GaAs and $Al_xGa_{1-x}As$ because of their close lattice matching [a(GaAs) = 5.653 Å, a(AlAs) = 5.662 Å] yet different energy gaps [E_g(GaAs) = 1.54 eV, E_g(AlAs) = 2.23 eV] and dielectric constants [n(GaAs) = 3.54, n(AlAs) = 2.96].

Optical waveguides are the simplest of the optical devices. They are used to transport information-bearing light energy between the optical components of an integrated-optic circuit by spatial confinement of the optical radiation field. The basic structure of a waveguide consists of a center guiding layer of refractive index n_1 surrounded by two cladding layers of refractive index n_2, where $n_1 \geqslant n_2$. For the fabrication of integrated optical circuits with GaAs lasers as light sources, waveguides with a material like $Al_xGa_{1-x}As$ having an index of refraction lower than that of GaAs are required, in order to couple light into the waveguide. Since the refractive index of $Al_xGa_{1-x}As$ decreases with increasing x, it is possible to build the desired waveguide simply by sheathing an appropriately shaped region of $Al_xGa_{1-x}As$ with another layer of larger x. Extremely smooth interfaces between cavity and cladding layers are important for an effective reduction of optical losses due to interface scattering.

Several authors[51, 193–195] have demonstrated that highly promising low loss $Al_xGa_{1-x}As$ passive optical waveguides can be fabricated by MBE. Using waveguides with x = 0.1, Merz and Cho[194] achieved optical losses of less than 1.5 cm^{-1} at the energy range of the GaAs laser emission. Similar results were obtained for waveguides with x = 0.25 in the central $Al_xGa_{1-x}As$ layer. Because of the highly reactive nature of $Al_xGa_{1-x}As$, a sufficient trapping of H_2O and CO by additional cryopanels is a stringent requirement during MBE growth.

A new approach to waveguiding in a slab dielectric bounded on one side by air and on the other by an MBE-grown periodic medium, which involves diffraction localization of the optical wave field, has recently been reported by Cho et al.[196]. The cladding layer of the waveguide with appropriate periodic variation in the refractive index consisted of alternating GaAs and $Al_{0.2}Ga_{0.8}As$ layers covering one or both sides of the $Al_{0.32}Ga_{0.68}As$ core. Interference between the optical radiation field and the periodic GaAs/$Al_{0.2}Ga_{0.8}As$ cladding layer exhibited an optical bandgap which confined the radiation field to the $Al_{0.32}Ga_{0.68}As$ core. These novel Bragg waveguides have a number of intriguing properties including optical confinement even when the core has a refractive index lower than that of the periodic layers and the possibility of being used for guiding in the medium to long X-ray region of the spectrum (this requires guides with periodic layers on both sides). Furthermore, such guiding should possess high wavelength selectivity.

Stimulated by these investigations, Yeh et al.[197] have experimentally observed single-mode optical surface-wave propagation along the surface of a periodic $GaAs/Al_{0.2}Ga_{0.8}As$ multilayer structure grown by MBE. This observation has clearly demonstrated the optical analogue of electronic surface states in crystals.

Monolithic integration of a double-heterostructure (DH) laser to a passive waveguide has been successfully achieved by *tapered optical couplers*. Extremely smooth gradual tapers were grown by MBE using knife-edge shadow masking[198]. For transfer of light from a GaAs to an $Al_{0.1}Ga_{0.9}As$ layer, coupling efficiencies approaching 100% have been observed. Reinhart and Cho[199] demonstrated the feasibility of integrating lasers with passive waveguides by fabricating taper-coupled lasers (TCL)[200] with MBE. The cross section of a representative structure of this type is schematically shown in Fig. 35. The sequence of layers for this complex structure was grown in one growth cycle by opening and closing the shutters and adjusting the temperatures of the appropriate effusion cells in the correct order. The tapered coupler was obtained by inserting a tantalum knife-edge in front of the substrate during growth of the active layer. Lasers with an active layer thickness of 0.3 μm fabricated separately from this structure exhibited a threshold current density, j_{th}, of 2.9 kA/cm^2 at room temperature. From the homogeneous passive $Al_yGa_{1-y}As$ waveguide section of the structure depicted in Fig. 35, MBE grown lasers oscillating with a wavelength of $\lambda \leqslant 0.83$ μm have been obtained for the first time. The fairly high threshold current density (j_{th} = 9 kA/cm^2) of these lasers approaching the visible region of the spectrum may possibly be reduced by further reduction of spurious H_2O and CO contamination during growth.

Room-temperature continuous-wave (CW) *double heterostructure* (DH) *injection lasers* represent the most important component for the development of an optical communication system. The best studied basic structure consists of a thin active GaAs layer sandwiched between two $Al_xGa_{1-x}As$ layers with $X \approx 0.3$. The charge carriers injected under forward bias as well as the optical radiation are confined to the active region which forms an optical resonant cavity in conjunction with the mirror faces. Cho et al.[7, 201, 202] have demonstrated that the fabrication of these

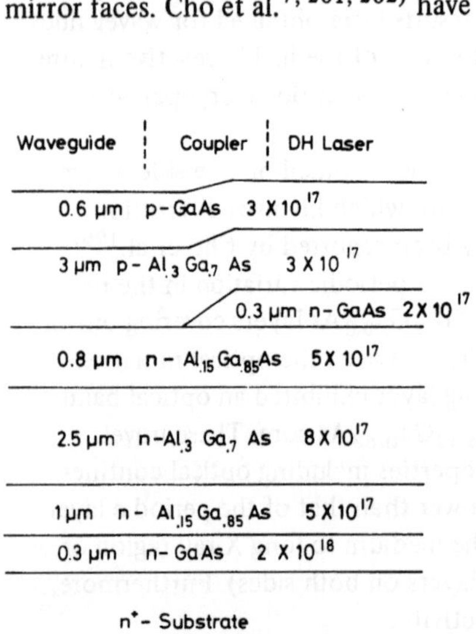

Fig. 35. Schematic layer sequence used for monolithic integration of a GaAs DH laser to a passive $Al_{0.1}Ga_{0.9}As$ waveguide by a tapered optical coupler

minority carrier devices imposes more stringent requirements on the quality of the epitaxial layers than for majority carrier devices described in Sect. 4.1.1. The performance of MBE grown DH lasers has been improved from room-temperature CW threshold current densities of 35 kA/cm^2 to 2.7 kA/cm^2 for 12 μm wide and 380 μm long cavities[139, 201, 202]. This considerable improvement was the result of several factors. First, an n-type active GaAs layer was chosen because of the higher photoluminescent intensity of n-type MBE GaAs as compared to p-type MBE GaAs and n-LPE GaAs[139]. Secondly, in order to improve the properties of the $Al_xGa_{1-x}As$ layers, a liquid-nitrogen-cooled panel was installed around the substrate for more effective pumping of H_2O and CO. Thirdly, the Al content was graded over a thickness of ~500 Å at all GaAs/$Al_xGa_{1-x}As$ interfaces in the structure. This was necessary to minimize the stress at the hetero-interface, since the GaAs/AlAs lattice parameter mismatch is significantly greater at the lower growth temperature of MBE ($T_S \approx 580$ °C) than at the 200 °C higher growth temperature of LPE. However, even this improved value of $j_{th} = 2.7$ kA/cm^2, obtained from a complex six-layer structure similar to that shown in Fig. 35, was still a factor of two above that value of similar-geometry lasers prepared by LPE. Tsang[203] recently proposed that bulk nonradiative recombination near the heterojunctions in the wide-band-gap regions may be an important contributor to the higher threshold currents of MBE grown DH lasers. Subsequently, Tsang[124] has then succeeded in growing GaAs/$Al_{0.3}Ga_{0.7}As$-DH lasers by MBE having threshold current densities ($j_{th} \leqslant 2$ kA/cm^2) as low as those of similar-geometry DH lasers grown by LPE. In addition, these broad-area lasers exhibited excellent stability and uniformity in lasing-filament distribution across the entire junction plane and throughout the entire current injection range. The differential quantum efficiencies were found to be about 40%. All values were obtained with 0.4 μm active layer thickness. Both Sn-doped n-type and Be-doped p-type active GaAs layers yielded comparable results. Three major improvements were applied during MBE growth; (i) using a sample load-lock system and additional cryopanels around the substrate; (ii) adjusting the on-off sequence of the Be and Sn dopant beams so that the p/n junction exactly matches the heterojunction (surface accumulation of Sn!); and (iii) increasing the growth temperature to $T_S \geqslant 600$ °C.

Pulsed room-temperature lasing at 1.65 μm of lattice-matched $Ga_{0.47}In_{0.53}As$/InP double-heterostructures grown by MBE has recently been demonstrated for the first time [204]. A threshhold current density as low as 3.2 kA/cm^2 for a 0.6 μm thick layer was achieved. This value compares favourably with the $Al_xGa_{1-x}As$ and with the $Ga_xIn_{1-x}P_yAs_{1-y}$ system.

Several other intriguing examples of MBE's capability to grow complex structures for both passive and active optoelectronic devices have been described in literature. Lee and Cho[205] applied the GaAs planar technology[50, 186] to fabricate a buried-stripe laser structure which exhibited single transverse optical-mode operation with an overall quantum efficiency of 20%. Casey et al.[206] used an MBE/LPE hybrid growth technique to fabricate a separate-confinement heterostructure (SCH) laser with periodic corrugation in the optical cavity for distributed feedback. A threshhold current density of 2.2 kA/cm^2 at room temperature was achieved for this structure. Cho et al. prepared back-surface emitting $GaAs_xSb_{1-x}$ LED's ($\lambda = 1.0$ μm)[141] and $Al_xGa_{1-x}As$/GaAs DH LED's[207] by MBE which exhibited promising diode characteristics. Finally, Ilegems et al.[208] have fabricated integrated structures consisting of two p-i-n double heterostructure GaAs photodetectors ohmically connected by an interposed tunnel diode, which yielded open-circuit voltages larger than the band gap of the absorption material and power efficiencies comparable to those of single-cell detectors.

4.1.3 Recent Developments

While in the field of MBE growth of GaAs and $Al_xGa_{1-x}As$ considerable progress has been made towards fabricating two-terminal optical devices as well as microwave devices with promising characteristics so that now straightforward improvements of device performance in the development area can be projected, in the research area the unique features of MBE have been further exploited in the production of (i) epitaxial structures with lateral dimensional controls by means of masking, (ii) metal-semiconductor heterojunctions, and (iii) insulator-semiconductor heterostructures.

A two-dimensional definition normal to the growth direction is required for many optoelectronic devices involving $GaAs/Al_xGa_{1-x}As$ multilayers. The *lateral dimensional control* during MBE has been achieved by various types of masking. In the approach using the MBE planar technology[50, 186] polycrystalline and monocrystalline GaAs were selectively grown producing a pattern which consisted of areas of device-quality single-crystalline material electrically isolated by semi-insulating polycrystalline material. The active device positions were defined by photolithographically etched windows in SiO_2 layers which covered the starting Cr-doped semi-insulating GaAs substrate. Single crystal areas of less than 5 μm on a side have been grown with this technique.

Another approach for planar definition involves growth of $GaAs/Al_xGa_{1-x}As$ layers over preferentially etched channels aligned along distinct directions on the substrate surface[122] (etch-and-fill technique). Figure 36a clearly demonstrates that for channels aligned along the $[\bar{1}\bar{1}0]$ direction of the (100) GaAs substrate, the two edges of the undercut channels serve as two edge shadowing masks preventing atoms from the incident flux arriving at the surface directly underneath. As a result, stripes of $GaAs/Al_xGa_{1-x}As$ multilayers are grown in the channel bound on all sides by single crystal planes completely disconnected from the rest of the epilayer. This effect arises from the unique growth characteristic of MBE over small surface areas in that the (100) planes are more favored than the (111) planes as indicated in Fig. 36b (for details see Sect. 2.3.1). Note that in Fig. 36 the various angles are distorted since the samples were tilted.

A third approach for two-dimensional definition permits the growth of mesa structures by using mechanical masks consisting of Si with patterned openings which is held fixed between the sources and the substrate. Using single- or multilevel masking, Tsang and Ilegems[52] fabricated single and multilayer stripe-mesa waveguides with widths as narrow as 1 μm, two-dimensional waveguide tapers and various other epilayer patterns with well-defined optically smooth mirrorlike edges. Tsang and Cho[53] have extended this technique and have achieved molecular beam epitaxial *writing* of patterned GaAs thin-film structures by moving the substrate behind the

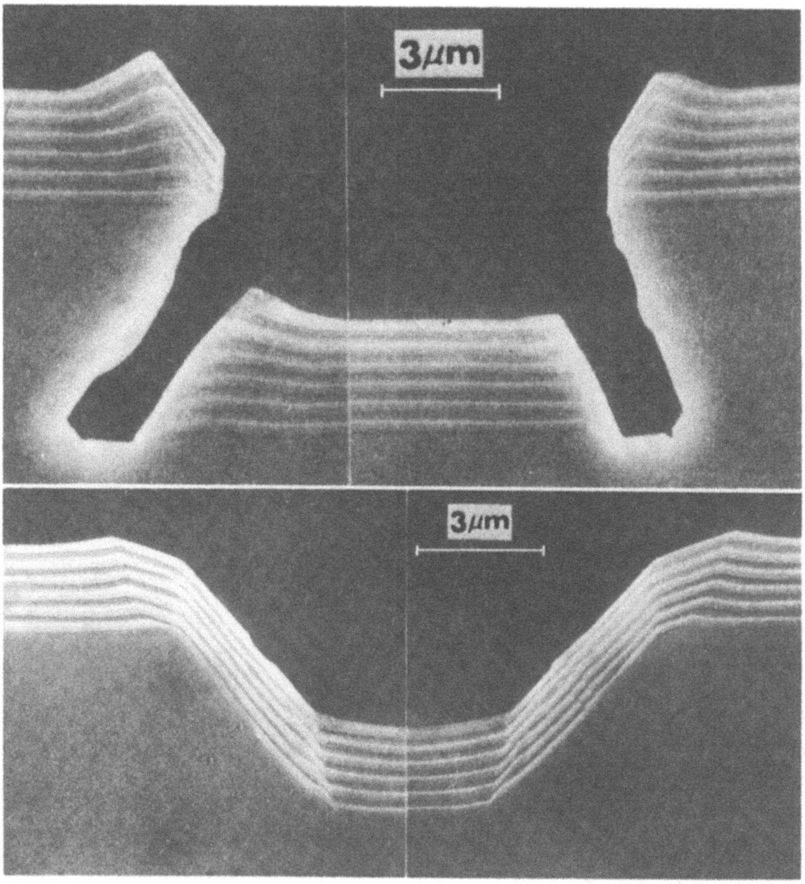

(a)

(b)

Fig. 36 a and b. SEM micrographs of the cross-sections of alternating GaAs (0.22 μm) and $Al_{0.2}Ga_{0.8}As$ (0.22 μm) multilayer structures grown over preferentially etched channels; **(a)** channel aligned along the $[\bar{1}10]$ direction; **(b)** channel aligned along the $[1\bar{1}0]$ direction. The samples were viewed at tilted angles resulting in a distortion of the various angles (from Ref.[122])

mechanical Si mask in a plane parallel to the mask. In addition, the thickness of the written epitaxial layer could be varied by varying either the velocity of the substrate motion during writing or by using mask openings of different sizes. The result of this last approach is demonstrated in Fig. 37. Using this technique, GaAs thin-film structures with controlled thickness profiles, lengthwise varying thickness, featureless and optically smooth sidewalls, and a linewidth as narrow as 10 μm have been written.

Low-resistance ohmic contacts to n- and p-type GaAs as well as Schottky barrier diodes on GaAs are of great technological importance for microwave and optoelectronic device fabrication. Recently, the unique capabilities of MBE have been used to grow first the epitaxial layer structure needed for device operation and then the required metal film onto the freshly deposited surface without breaking the

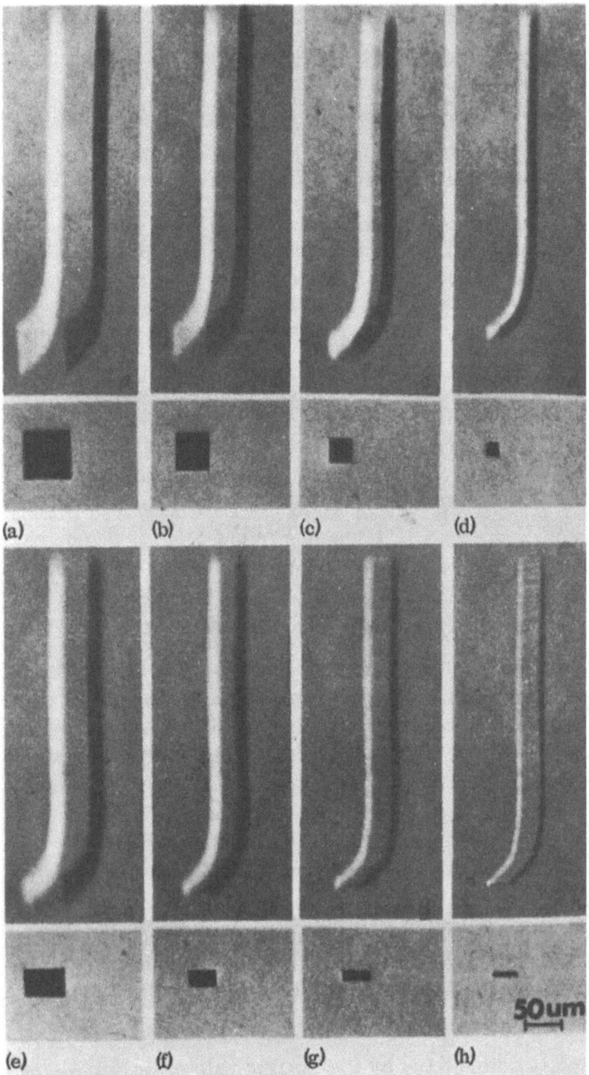

Fig. 37. GaAs epitaxial layer structures *written* behind a series of square and rectangular mask openings having different sizes. The mask openings corresponding to each of the structures are shown below the epitaxial layer structures (from Ref. [53])

vacuum in the same growth cycle. During this procedure the semiconductor surface is never exposed to the ambient air and to solvants which may leave undesired contaminants on the surface before plating. Thus, MBE in-situ metallization has made feasible the reproducible realization of ideal *semiconductor-metal interfaces.* As described in Sect. 3.2, Cho and Dernier[165] have obtained Schottky barrier diodes formed by epitaxially grown monocrystalline Al on (100) GaAs, which exhibited nearly ideal electrical characteristics. The 3000 Å thick Al layers were grown at room temperature in order to avoid the onset of three-dimensional growth. In this case the Al films were (100) oriented on the (100) GaAs surface with a 45° rotation around the [100] axis, in order to accomodate the difference in their lattice constants. The measured barrier heights depended significantly on the surface stoichiometry of the active GaAs layer (cf. Sect. 3.2). The ideality factor of such diodes was found to be approximately 1.04.

The formation of ohmic contacts to both n- and p-type GaAs by MBE without the need for a high-temperature alloying has recently been demonstrated by several authors. In these in-situ metallization experiments, either Sn[114, 209)], Ge[61, 210)] or Be[211)] was deposited on freshly grown n^{++}- or p^{++}-GaAs layers, respectively, doped with the corresponding dopant element. Subsequently, Au/Ge, Au/Mo, or Au/Zn ohmic contacts, respectively, could be readily formed on these highly doped layers without exceeding the eutectic temperature. These non-alloyed ohmic contacts exhibited linear current-voltage characteristics and yielded specific contact resistance in the order of 1×10^{-6} Ω cm^{-2}. Power GaAs FET's fabricated from these structures showed excellent rf and dc characteristics[209)].

Dielectric films on GaAs for MOS device application have been prepared by various methods. However, although these films display the required electrical characteristics, they often exhibit high surface state densities and fixed charge densities at the *semiconductor-dielectric interface*, along with variable charge injection from the semiconductor. New approaches to the fabrication of metal/insulator/GaAs heterostructures based on the use of either insulating oxygen-doped $Al_xGa_{1-x}As$ or amorphous aluminium oxide films grown in-situ by molecular beam reaction under UHV conditions have now been established. Casey et al.[212–214)] produced lattice matched single-crystal insulator-semiconductor heterojunctions consisting of a 2000 Å thick high-resistivity oxygen-doped $Al_{0.5}Ga_{0.5}$ As layer deposited on an n-GaAs layer grown in the same cycle. DLTS and C-V measurements on such n-GaAs-$Al_{0.5}Ga_{0.5}$As-Cr/Au MIS capacitors revealed the absence of significant interface traps in these structures and showed that the heterojunction was near flat band at zero bias. Also, a significant reduction of the surface recombination current in Zn-diffused GaAs p/n junctions by using an oxygen-doped $Al_{0.5}Ga_{0.5}As$ surface layer has recently been demonstrated[214)].

Another approach[23, 25)] for application in GaAs MOS capacitors used MBE grown a-Al_2O_3/$Al_xGa_{1-x}As$/GaAs heterostructures. C-V measurements on Al/a-Al_2O_3/$Al_{0.33}Ga_{0.67}$As/n-GaAs MOS capacitors have shown that favorable mechanical and electrical behavior can be obtained with amorphous Al_2O_3 layers grown by the reaction of a thermal Al molecular beam with oxygen gas at low ($\leqslant 250$ °C) temperatures. A passivation of the as-grown GaAs surface by a thin $Al_xGa_{1-x}As$ layer was found to be effective in preserving the GaAs stoichiometry during growth of the Al_2O_3 layer, resulting in a low density of interface traps. In the Raman spectrum of an a-Al_2O_3/$Al_xGa_{1-x}As$/GaAs heterostructure shown in Fig. 30 the vibrational spectra of each individual layer can be clearly resolved. Additional peaks arising from the amorphous Al_2O_3 layer or caused by any undefined species resulting from the oxidation process were never detected in these Raman investigations[25)].

The few selected examples have shown that the MBE technique may provide a number of novel processing steps ideally suited for standard and more complex device structures. In addition, the recent developments of MBE clearly demonstrate that the limits of this technique for exploration of novel device structures have not yet been reached.

4.2 Research Application of MBE

In the preceeding sections it has been demonstrated that the MBE technique enables us to prepare flat, well-ordered, clean semiconductor surfaces of distinct orientation and to alter the surface composition in a well-defined reproducible manner. Both features are essential for sophisticated surface experiments including AES, LEED, LEELS, XPS, and UPS as well as Raman scattering. A brief outline of some possible applications of these experiments to in situ MBE-grown material has been given in a recent publication[75].

The devices discussed in Sect. 4.1 require structural dimensions which are large compared to the lattice dimensions. With the growth of ultra-thin multilayer structures (see Figure 40), however, MBE has demonstrated its unique capability for growing well-defined structures exhibiting dimensions of the same order as the lattice spacing in the direction of growth. Historically, in 1970, Esaki and Tsu[215] proposed that a compound semiconductor heterostructure having a one-dimensional periodic variation of alloy composition, referred to as a "superlattice", where the period should be shorter than the electron mean free path, would exhibit quite unusual, novel transport properties. The striking properties of a semiconductor superlattice should arise from the formation of subbands due to a division of the Brillouin zone into a series of minizones. Subsequently, multilayer structures consisting of GaAs

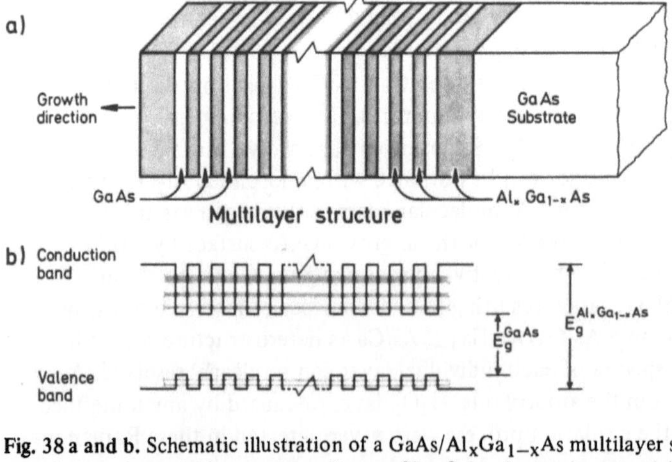

Fig. 38 a and b. Schematic illustration of a $GaAs/Al_xGa_{1-x}As$ multilayer structure of periodicity *d*. Also shown is the corresponding profile of the electronic potential energy

layers of 50 Å to 500 Å thickness interleaved with $Al_xGa_{1-x}As$ (x = 0.20–0.35) layers were successfully produced by molecular beam epitaxy[5, 95, 216] in which there were typically 100 periods. Some of the predicted intriguing phenomena of quantum confinement of electron waves could be verified experimentally[217–222]. Figure 38 shows a schematic illustration of a $GaAs/Al_xGa_{1-x}As$ multilayer structure which may contain up to several hundred ultrathin layers. Also shown in the corresponding profile of electronic potential energy. Since the band gap of $Al_xGa_{1-x}As$ is greater than that of GaAs, the forbidden gap between valence and conduction band edges will go through sharp steps at each $GaAs/Al_xGa_{1-x}As$ interface. Experimental work has confirmed that electrons confined in the GaAs layers experience abrupt potential barriers with rise and fall distances of less than 5 or 10 Å[221]. It should be noted that the energy discontinuities of the valence and conduction bands are 15 ± 3% and 85 ± 3%, respectively, of the total energy gap difference in the Al-Ga-As system[219, 221]. Since a superlattice period of 50 to 100 Å is still large compared to the lattice constant, the one-dimensional periodic potential gives rise to quantum subbands of two-dimensional character separated by forbidden energy gaps. The energy positions and bandwidths of these subbands are determined by the barrier ($Al_xGa_{1-x}As$) and well (GaAs) thickness and by the barrier height, i.e., by the value of x. One of the major objections in the more recent studies of superlattices has been the experimental elucidation of the two-dimensional character of these electronic subbands by means of Raman scattering[223, 224] and Shubnikov-de Haas-effect measurements[225–227].

In the system $GaAs/Al_xGa_{1-x}As$, the conduction band edges are sufficiently far in energy from the valence band edges. Recently, Sai-Halasz et al.[228] have proposed a new semiconductor superlattice consisting of periodic $In_{1-x}Ga_xAs/GaSb_{1-y}As_y$ multilayers, in which the bottom of the conduction band of $In_{1-x}Ga_xAs$ is located very close to or even below the top of the valence band of $GaSb_{1-y}As_y$, depending on the compositions x and y. In the systems InAs/GaSb the bottom of the conduction band of InAs lies in fact approximately 0.15 eV *below* the top of the valence band of GaSb[229]. Thus, a strong interaction between the two bands exists, which should generate unusual electronic properties not observed in $GaAs/Al_xGa_{1-x}As$. The formation of distinct two-dimensional electronic subband structures in InAs/GaSb and $In_{1-x}Ga_xAs/GaSb_{1-y}As_y$ superlattices grown by MBE has been experimentally manifested by Shubnikov-de Haas oscillations[230] and by optical absorption measurements[229]. Furthermore, LCAO calculations for the InAs/GaSb surperlattice[231] led to the conclusion that this structure behaves like a semiconductor when the layers of the constituent material are thin ($\leqslant$ 100 Å), whereas the conduction and valence bands cross when the single layer thickness is increased to over ~ 115 Å resulting in a semimetallic behaviour. This semiconductor-semimetal transition in InAs/GaSb superlattices has recently been observed in Shubnikov-de Haas-effect measurements[232].

Large-angle X-ray scattering measurements[233, 235] confirmed that the individual layers in $GaAs / Al_xGa_{1-x}As$ multilayer structures are uniform and flat on an atomic scale with abrupt composition profiles within 5 Å in the direction of growth. Interdiffusion of Ga/Al as determined by Auger profiling[84] and optical absorption spectroscopy[234] as a function of annealing temperature was found to be surprisingly low. At typical growth temperatures of ~ 580 °C, the diffusion lengths are well below one monolayer assuming a deposition time of 3 h.

Gossard et al.[6, 41, 236] have recently exploited all unique features of the MBE growth process described in the proceeding sections in successfully producing ordered layered crystals of $(GaAs)_n$ and $(AlAs)_m$ repeated many thousands of times by alternate monolayer deposition. The number of deposited (GaAs) atomic layers, n, and of (AlAs) layers, m, were adjusted by a system of synchronously rotating shutters in front of the Ga and Al effusion cells. A model of a section of such an ideal $(GaAs)_n$ $(AlAs)_m$ alternate monolayer crystal with a (1.1) sequence grown on a (100) substrate is schematically shown in Fig. 39. Transmission electron microscopy (TEM) of cleaved and thinned samples has turned out to be a severe test for the achievement of the desired monolayer structures[41, 237]. Figure 40 shows a dark field TEM micrograph of a $(GaAs)_4$–$(AlAs_4)$ monolayer crystal confirming a well-ordered superlattice. It should be pointed out that the transmission electron diffraction pattern corresponds to a selected area of the probed sample of about 7000–8000 Å diameter. Thus, Fig. 40 indicates a continuous long-range ordering in the direction of growth for this $(GaAs)_4(AlAs)_4$ monolayer structure which was grown at 580 °C.

The alternate monolayer structures should be regarded as a novel class of "man-made" semiconductor materials exhibiting unique electronic properties which are probably unobservable in any "natural" crystal. The cubic T_d symmetry of the original materials is reduced to the tetragonal D_{2d} symmetry of the artificial material. In these structures, electrons in adjacent layers interact strongly, and the wave functions extend across several periods. Complete new band structure calculations were described by several authors[238–240] for (100)-oriented GaAs/AlAs monolayer alloys. In order to elucidate the optical properties as well as phonon properties and zone folding effects of these novel structures, various experimental studies have been performed[241–244].

The concept of alternate monolayer growth by MBE has been successfully extended to other semiconductor systems, in which the lattices of the constituent com-

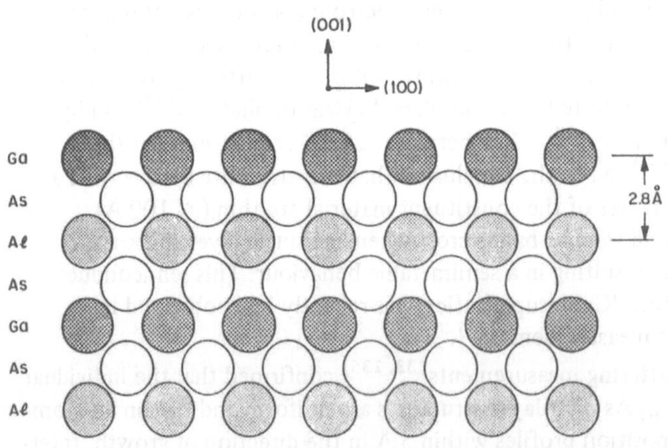

Fig. 39. Schematic atomic arrangement of an ideal $(GaAs)_1(AlAs)_1$ alternate monolayer crystal grown in (100) orientation (from Ref. [236])

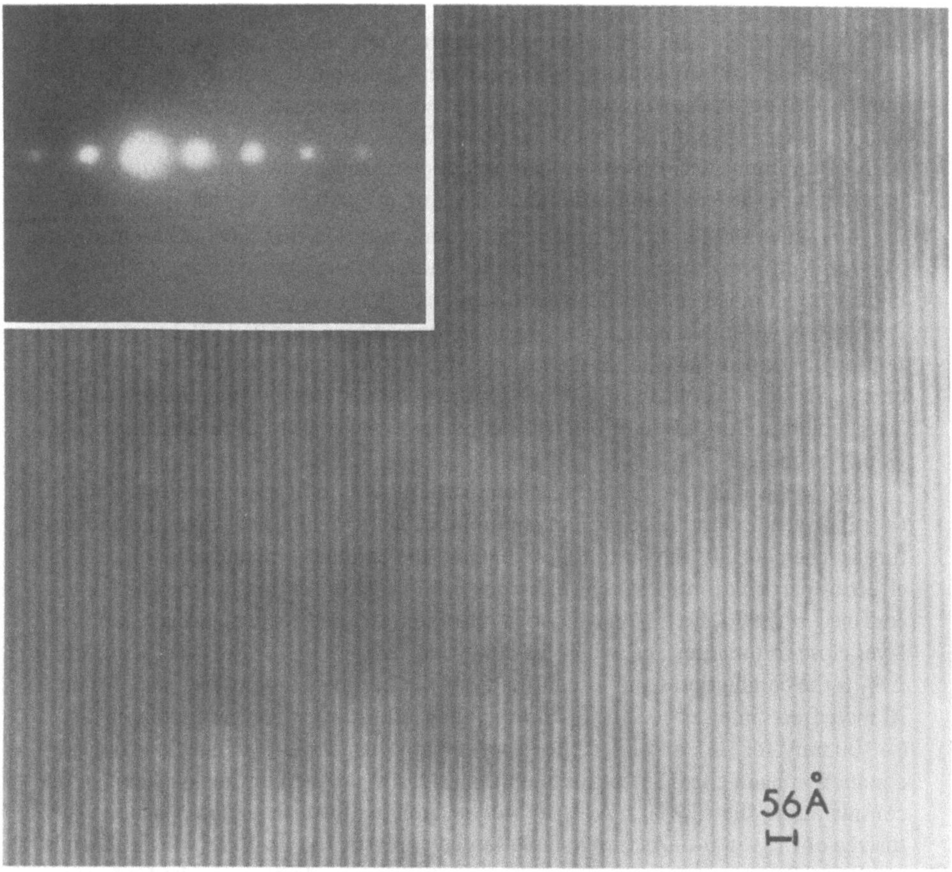

Fig. 40. Dark field transmission electron micrograph of a (110) cleaved and thinned $(GaAs)_4(AlAs)_4$ monolayer crystal grown at $T_S = 580\,°C$ indicating a well ordered superlattice (from Ref. [41])

pounds match closely. In the system $(Al_xGa_{1-x}As)_n$–$(Ge_2)_m$, smooth (100) interfaces were produced for x = 1 down to n = m = 7 at growth temperatures below 500 °C, although stacking faults were still present, whereas for low x and growth temperatures above 500 °C epitaxial columnar structures dominate with all n and m values[245]. With the $(GaAs)_n$-$(InAs)_m$ system, an ordered (100) monolayer structure was verified at a growth temperature of 400 °C[246]. A comprehensive review on all these intriguing properties of both semiconductor superlattices and monolayer structures will be presented in the near future by Gossard[247].

Although several suggestions have been made, the practical application of semiconductor structures exhibiting quantum size effects, i.e. superlattices, in actual device structures has been quite limited. This is partly due to the relatively poor carrier mobilities found in uniformly doped structures, and partly due to problems arising from the lattice mismatch at the numerous heterojunctions contained in these structures. Only very recently, GaAs/$Al_xGa_{1-x}As$ superlattices have been grown by MBE[153] in which both low- and room-temperature electron mobilities are significantly higher than those in otherwise equivalent epitaxial GaAs grown by other techniques. This new behaviour was achieved by synchronizing the change of composition with a change in doping, so that only the $Al_xGa_{1-x}As$ layers are deliberately doped with donor impurities. However, the production of these "modulation-doped" heterojunction superlattices is still rather laborious for any potential device application.

Abstreiter and Ploog[154] have recently demonstrated that most of the intriguing electronic properties expected for modulation doped superlattices can also be obtained from a much simpler heterostructure, consisting only of a thick (~ 1 μm) deliberately Ge-doped n-$Al_xGa_{1-x}As$ layer covered by a thin (0.1–0.2 μm) not intentionally doped GaAs layer grown by MBE in one straightforward growth cycle on a (100) GaAs substrate (Fig. 41 a). This structure having GaAs as the top epitaxial layer facilitates formation of electrical contacts as opposed to other structures previously proposed[155].

MBE-grown $Al_xGa_{1-x}As$/GaAs heterojunctions exhibit abrupt steps in conduction and valence band edges at a common interface[221]. This behaviour in band structure is schematically shown in Fig. 41 b. Only the conduction band is involved in the effect of particular interest. The electrons from those donor impurities which are confined in the $Al_xGa_{1-x}As$ will move into the GaAs and form an accumulation layer close to the interface, when the condition $(\Delta E_C - E_F) \geqslant E_D$ is fulfilled, where ΔE_C is the conduction band discontinuity at the interface (about 200 meV for an Al mole fraction of x = 0.20), E_D is the donor ionization energy of the impurities ($\leqslant$ 100 meV for Ge), and E_F is the Fermi energy in bulk GaAs measured from the conduction band edge. Since this condition is easily satisfied in the structures under consideration, the $Al_xGa_{1-x}As$ layer will be depleted towards the interface. At the surface of the top n-GaAs film a depletion layer exists, the depth of which depends on the (bulk) carrier concentration. As a result, the free electrons in the GaAs are strongly confined into an approximately triangular, one-dimensional potential well perpendicular to the interface similar to an accumulation layer in metal-oxide-semi-

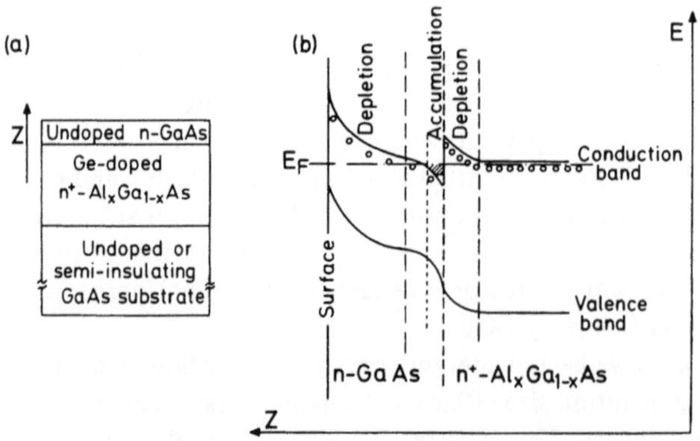

Fig. 41. (a) Geometry of abruptly doped n-GaAs/n^+-$Al_xGa_{1-x}As$ samples; (b) Energy band diagram of abruptly doped n-GaAs/n^+-$Al_xGa_{1-x}As$ heterojunction

conductor structures. Carrier motion along the interface is unhindered and practically free of ionized-impurity scattering because the parent donor impurities and the free electrons are spatially separated. However, the electronic motion perpendicular to the interface is quantized forming a set of bound states. Thus, a quasi-two-dimensional Fermi gas of high mobility results.

Direct evidence that a two-dimensional Fermi gas does exist in these structures was obtained from measurements of the field orientation dependence of Shubnikov-de Haas oscillations[248)]. Using Raman spectroscopy, Abstreiter and Ploog[154)] at first obtained inelastic light scattering under resonance conditions from a quasi-two-dimensional electron gas confined at the interface of abruptly doped GaAs/n-$Al_xGa_{1-x}As$ heterojunctions. For the first time intersubband excitations in a two-dimensional electron system were observed by this technique. Moreover, the two-dimensional electron concentration could be controlled by applying a voltage across a Schottky barrier fabricated on the wafer. With increasing reverse bias the depletion width increases, and finally the accumulation layer at the heterointerface was also depleted.

This quasi-two-dimensional electron gas bound to the interface of a semiconductor-semiconductor heterojunction may offer a highly promising system to study various intriguing electronic properties relevant to fundamental physics. In addition, these systems are of considerable technological importance for application in devices, for example in Schottky-barrier FET's[249)], because of their significantly enhanced electron mobilities.

5 Concluding Remarks

At this time of continuing rapid development and sophistication in crystal and thin film growth techniques the basic aspects of a new technique, molecular beam epitaxy (MBE), have been described and its unique capabilities for large-area uniform growth of III–V compound semiconductors with improved control over microscopic and even atomic multilayer dimensions have been reviewed.

Molecular beam epitaxy has emerged as a practical growth technique from its origin as a technique for surface studies of the interaction of thermal molecular beams with solid surfaces under UHV conditions. A number of surface-analytical techniques have been applied continuously to assure that the intended conditions have been reached before and are being maintained during growth. The results obtained from surface-analytical studies combined with surface-kinetic measurements led to a detailed insight into the growth mechanism during molecular beam epitaxy. Despite some sophistication in instrumentation, MBE has now achieved an accepted position among the more conventional growth processes LPE and VPE, and the MBE technique is diffusing rapidly to many laboratories throughout the world.

In the first decade, the fundamental process of MBE was developed mainly in industrial research laboratories (Bell Laboratories, IBM Watson Research Center, Philips Research Laboratories). During this period emphasis was on the MBE production of standard device structures exhibiting characteristics comparable or even superior to those fabricated by the more conventional growth and processing techniques. Excellent results were obtained for MBE grown GaAs microwave and optoelectronic devices including varactor diodes, mixer diodes, IMPATT diodes, MESFET's, optical waveguides and DH injection lasers. The unique features of MBE have since been further exploited to more complex device structures involving novel processing steps such as lateral dimensional control as well as metal-semiconductor and oxide-semiconductor heterojunction formation in a single growth cycle. For this exploration of novel device structures the limits of the MBE technique have not yet been reached.

The microwave and optoelectronic devices mentioned above require structural dimensions large in comparison with the lattice dimensions. A particularly attractive feature of MBE is the possibility it provides of growing multilayer structures exhibiting dimensions of the same order as the lattice spacing or even atomic dimensions in the direction of growth. These "man-made" quantum-well and monolayer structures consist of hundreds of alternating layers of different material (e.g., GaAs and AlAs or InAs and GaSb). They display intriguing novel electronic and structural features which are probably unobservable in any "natural" crystal. Recently, even intersubband excitations have been observed in a quasi-two-dimensional electron gas confined at the interface of a semiconductor-semiconductor heterojunction grown by molecular beam epitaxy.

In solid-state device technology there is a growing demand for devices exhibiting smaller dimensions, higher frequency and greater efficiency. Therefore, one of the most significant functions of modern MBE will be that of being a driving force for continuous refinement of the more conventional growth techniques. Complex semiconductor structures, which serve as model structures, can now be produced by MBE. These structures may either be applied in novel devices with improved characteristics or which may be used to study fundamental physical properties. If the technological importance of such MBE grown, novel semiconductor structures becomes established, attempts can subsequently be made to grow the same or similar structures by less sophisticated techniques. Several examples for this motivating function of MBE have recently been published in the literature. First, Dupuis et al.[251, 252] prepared high-quality single and multiple GaAs/$Al_xGa_{1-x}As$ quantum-well heterostructures by means of metalorganic chemical vapour deposition (MO–CVD). Second, Tsui and Logan[156] used an improved LPE technique to grow abruptly doped GaAs/n-$Al_xGa_{1-x}As$ heterojunctions with interfacial electrical properties comparable to those achieved by MBE[154].

A general trend in instrumentation is towards less sophisticated and more user-oriented MBE systems involving substrate exchange load-lock systems, extended cryopanels within the growth chamber, and automatation of molecular beam generation and control. Still an unresolved problem is the reproducibility and reliability of element-specific molecular beam detectors for closed-loop computer controlled MBE systems. Another significant trend in commercial manufacturing of MBE systems is the increasing use of the mechanical assembly technique, which greatly facilitates the purchase of modern MBE systems.

So far, the MBE technique has mainly been used to produce thin films of III–V, II–VI, and IV–VI compounds and group IV elements (Si, Ge, etc.) which may also be prepared by other methods. The unique features of molecular beam epitaxy, however, permit also the growth of stable or even metastable solid compounds of novel elemental composition in kinetically controlled reactions. Therefore, the application of the MBE technique in the field of *preparative solid-state chemistry* may well become one of the new highlights in the near future.

Acknowledgment. The author is indebted to the many of his colleagues working in the field of MBE growth for providing preprints and reprints and for the permission to reproduce various figures, as noted in the captions. The expert help in sample preparation of A. Fischer and in sample characterization of H. Künzel and Miss H. Willerscheid is gratefully acknowledged. The author would also like to express his appreciation to H. J. Queisser and to G. B. Stringfellow for critical reading of the manuscript, and to Ms. I. Zane for typing the manuscript. The work was supported by the Bundesministerium für Forschung und Technologie of the Federal Republic of Germany.

6 References

1. Cho, A. Y., Arthur, J. R.: Progr. Solid State Chem. *10*, 157 (1975)
2. Günther, K. G.: Z. Naturforsch. *13a*, 1081 (1958)
3. Davey, J. E., Pankey, T.: J. Appl. Phys. *39*, 1941 (1968)
4. Arthur, J. R.: J. Appl. Phys. *39*, 4032 (1968)
5. Chang, L. L., Esaki, L., Howard, W. E., Ludeke, R.: J. Vac. Sci. Technol. *10*, 11 (1973)
6. Gossard, A. C., Petroff, P. M., Wiegmann, W., Dingle, R., Savage, A.: Appl. Phys. Lett. 323 (1976)
7. Cho, A. Y.: Japan. J. Appl. Phys. *16* Suppl. 16-1, p. 435 (1977)
8. Walpole, J. N., Calawa, A. R., Harman, T. C., Groves, S. H.: Appl. Phys. Lett. *28*, 552 (1976)
9. Joyce, B. A., Foxon, C. T.: Proc. VI. ESSDERC 1976, Inst. Phys. Conf. Ser. *32*, 17 (1977)
10. Chang, L. L., Ludeke, R.: in Epitaxial growth, p. 37. J. W. Matthews ed. New York: Academic 1975
11. Farrow, R. F. C.: In: 1976 crystal growth and materials, Vol. 1, p. 237. E. Kaldis and H. J. Scheel, eds., Amsterdam: North Holland 1977
12. Smith, D. L., Pickhardt, V. Y. J. Electrochem. Soc. *125*, 2042 (1978)
13. Smith, D. L.: Prog. Crystal Growth Charact., *2*, 33 (1979)
14. Knudsen, M.: Ann. Phys. [4] *28*, 999 (1909)
15. Clausing, P.: Ann. Phys. [5] *12*, 961 (1932)
16. Ruth, V., Hirth, J. P.: In: Condensation and evaporation of solids, p. 99. E. Rutner, P. Goldfinger, and J. P. Hirth eds. New York: Academic 1964
17. Arthur, J. R.: In: Proc. conf. structure and chem. solid surfaces, p. 46–1. G. A. Somorjai ed. New York: Wiley 1967
18. Arthur, J. R., LePore, J. J.: J. Vac. Sci. Technol. *6*, 545 (1969)
19. Foxon, C. T., Joyce, B. A.: J. Crystal Growth *44*, 75 (1978)
20. Ploog, K., Fischer, A.: Appl. Phys. *13*, 111 (1977)
21. Fischer, A., Graf, K., Hafendörfer, M., Künzel, H., Ploog, K.: J. Phys. (E), to be published (1980)
22. Wood, C. E. C., Joyce, B. A.: J. Appl. Phys. *49*, 4854 (1978)
23. Hirose, M., Fischer, A. and Ploog, K.: Phys. Stat. Sol. (a) *45*, K175 (1978)
24. Becker, G. E.: J. Vac. Sci. Technol. *14*, 640 (1977)
25. Ploog, K., Fischer, A., Trommer, R., Hirose, M.: J. Vac. Sci. Technol. *16*, 290 (1979)
26. Foxon, C. T., Joyce, B. A.: Private communication (1978)
27. Foxon, C. T., Boudry, M. R., Joyce, B. A.: Surf. Sci. *44*, 69 (1974)

28. Farrow, R. F. C.: J. Phys. *D 7,* 2436 (1974)
29. Ilegems, M., Dingle, R., Rupp, L. W., Jr.: J. Appl. Phys. *46,* 3059 (1975)
30. Ilegems, M.: J. Appl. Phys. *48,* 1278 (1977)
31. Shen, L. Y. L.: J. Vac. Sci. Technol. *15,* 10 (1978)
32. Chatillon, C., Allibert, M., Pattoret, A.: Adv. Mass. Spectr. *7A* (1978) 615
33. Foxon, C. T.: Acta Electronica *16,* 323 (1973)
34. Smith, D. L., Pickhardt, V. Y.: J. Appl. Phys. *46,* 2366 (1975)
35. Holloway, H., Hohnke, D. K., Crawley, R. L., Wilkes, E.: J. Vac. Sci. Technol. *7,* 586 (1970)
36. Nagata, S., Tanaka, T., Fukai, M.: Appl. Phys. Lett. *30,* 503 (1977)
37. Neave, J. H., Joyce, B. A.: J. Cryst. Growth *44,* 387 (1978)
38. Foxon, C. T., Harvey, J. A., Joyce, B. A.: J. Phys. Chem. Solids *34,* 1693 (1973)
39. Esaki, L.: Proc. 6th Intern. Vac. Congr. 1974, Japan. J. Appl. Phys. Suppl. *2,* Pt. *1,* 821 (1974)
40. Etienne, P., Massies, J., Linh, N. T.: J. Phys. *E10,* 1153 (1977)
41. Petroff, P. M., Gossard, A. C., Wiegmann, W., Savage, A.: J. Crystal Growth *44,* 5 (1978)
42. Wood, C. E. C.: Appl. Phys. Lett. *29,* 746 (1976)
43. Kometani, T. Y., Wiegmann, W.: J. Vac. Sci. Technol. *12,* 933 (1975)
44. Sentinel 200 is a product of Inficon Leybold-Heraeus Inc., E. Syracuse, N. Y. 13057, U.S.A.
45. Brewer, L., Kane, J. S.: J. Phys. Chem. *59,* 105 (1955)
46. Matsushima, V., Hirofuji, Y., Gonda, S., Mukai, S., Kimata, M.: Japan. J. Appl. Phys. *15,* 2321 (1976)
47. Arthur, J. R.: J. Phys. Chem. Solids *28,* 2257 (1967)
48. Foxon, C. T., Joyce, B. A., Farrow, R. F. C., Griffiths, R. M.: J. Phys. *D 7,* 2422 (1974)
49. Cho, A. Y., Reinhart, F. K.: J. Appl. Phys. *45,* 1812 (1974)
50. Cho, A. Y., Ballamy, W. C.: J. Appl. Phys. *46,* 783 (1975)
51. Cho, A. Y., Reinhart, F. K.: Appl. Phys. Lett. *21,* 355 (1972)
52. Tsang, W. T., Ilegems, M.: Appl. Phys. Lett. *31,* 301 (1977)
53. Tsang, W. T., Cho, A. Y.: Appl. Phys. Lett. *32,* 491 (1978)
54. Cho, A. Y.: J. Appl. Phys. *41,* 2780 (1970)
55. Cho, A. Y., Panish, M. B., Hayashi, I.: Proc. 3rd Int. Symp. on GaAs and Related Compounds, Aachen, 1970, Inst. Phys. Conf. Ser. *9,* 18 (1971)
56. Cho, A. Y., Hayashi, I.: Solid-State Electron. *14,* 125 (1971)
57. For alternate solutions, which are effective during mechanochemical etching, see E. W. Jensen, Solid State Technol. *16,* 49 (1973)
58. Cho, A. Y., Panish, M. B.: J. Appl. Phys. *43,* 5118 (1972)
59. Arthur, J. R.: J. Appl. Phys. *38,* 4023 (1967)
60. Cho, A. Y.: J. Appl. Phys. *46,* 1733 (1975)
61. Wood, C. E. C., Woodcock, J., Harris, J. J.: Proc. 7th Int. Symp. on GaAs and Related Compounds, St. Louis, 1978, Inst. Phys. Conf. Ser. *45,* 28 (1979)
62. Matysik, K. J: J. Appl. Phys. *47,* 3826 (1976)
63. Delhomme, B. J., Blanchet, R. C., Fumey, M., Urgell, J. J.: Preprint (1979) submitted for publication to Rev. Phys. Appl.
64. Neave, J. H., Joyce, B. A.: J. Crystal Growth *43,* 204 (1978)
65. Abstreiter, G., Bauser, E., Fischer, A., Ploog, K.: Appl. Phys. *16,* 345 (1978)
66. Laurence, G., Simondet, F., Saget, P.: Appl. Phys. *19,* 63 (1979)
67. Cho, A. Y., DiLorenzo, J. V., Hewitt, B. S., Niehaus, W. C., Schlosser, W. O., Radice, C.: J. Appl. Phys. *48,* 346 (1977)
68. McFee, J. H., Miller, B. I., Bachmann, K. J.: J. Electrochem. Soc. *124,* 259 (1977)
69. Asahi, H., Okamoto, H., Ikeda, M., Kawamura, Y.: Japan. J. Appl. Phys. *18,* 565 (1979)
70. Farrow, R. F. C.: J. Phys. *D 7,* L121 (1974)
71. Arthur, J. R.: Surf. Sci. *43,* 449 (1974)
72. Foxon, C. T., Joyce, B. A.: Surf. Sci. *50,* 434 (1975)
73. Joyce, B. A., Foxon, C. T.: J. Crystal Growth *31,* 122 (1975)
74. Foxon, C. T., Joyce, B. A.: Surf. Sci. *64,* 293 (1977)
75. Ploog, K.: J. Vac. Sci. Technol. *16,* 838 (1979)

76. Chang, C. C.: In: Characterization of solid surfaces, p. 509. P. F. Kane and G. B. Larrabee eds. New York: Plenum Press 1974
77. LeGressus, C., Massignon, D., Sopizet, R.: C. R. Acad. Sci. (Paris), Ser. B, *280*, 439 (1975)
78. Ploog, K., Fischer, A.: J. Vac. Sci. Technol. *15*, 255 (1978)
79. Hammer, R., Chou, N. J., Eldridge, J. M.: J. Electron. Mater. *5*, 557 (1976)
80. Arthur, J. R., LePore, J. J.: J. Vac. Sci. Technol. *14*, 979 (1977)
81. Dingle, R., Logan, R. A., Arthur, J. R.: Proc. 6th Int. Symp. on GaAs and Related Compounds, Edinburgh, 1976, Inst. Phys. Conf. Ser. *33a*, 210 (1977)
82. Ludeke, R., Esaki, L., Chang, L. L.: Appl. Phys. Lett. *24*, 417 (1974)
83. Coburn, J. W., Kay, E.: CRC Crit. Rev. Solid State Sci. *4*, 561 (1974)
84. Chang, L. L., Koma, A.: Appl. Phys. Lett. *29*, 138 (1976)
85. Ludeke, R., Esaki, L.: Surf. Sci. *47*, 132 (1975)
86. Ludeke, R., Koma, A.: CRC Crit. Rev. Solid State Sci. *5*, 259 (1975)
87. Ludeke, R., Koma, A.: J. Vac. Sci. Technol. *13*, 241 (1976)
88. Bauer, E.: In: Techniques of metals research, Vol. III, part 2, p. 501. R. F. Bunshah, ed. New York: Wiley 1969
89. Cho, A. Y.: J. Appl. Phys. *42*, 2074 (1971)
90. Menadue, J. F.: Acta Cryst. *A28*, 1 (1972)
91. Dove, D. B., Ludeke, R., Chang, L. L.: J. Appl. Phys. *44*, 1897 (1973)
92. Masud, N., Pendry, J. B.: J. Phys. *C9*, 814 (1976)
93. Holloway, S., Beeby, J. L.: J. Phys. *C11*, L247 (1978)
94. Jona, F.: IBM J. Res. Dev. *9*, 375 (1965)
95. Chang, L. L., Esaki, L., Howard, W. E., Ludeke, R., Schul, G.: J. Vac. Sci. Technol. *10*, 655 (1973)
96. Cho, A. Y.: J. Appl. Phys. *47*, 2841 (1976)
97. Ludeke, R.: IBM J. Res. Dev. *22*, 304 (1978)
98. Joyce, B. A., Neave, J. H., Foxon, C. T.: Private communication (1978)
99. Cho, A. Y., Hayashi, I.: J. Appl. Phys. *42*, 4422 (1971)
100. Ploog, K., Fischer, A., Künzel, H.: Appl. Phys. *18*, 353 (1979)
101. Appelbaum, J. A., Baraff, G. A., Hamann, D. R.: J. Vac. Sci. Technol. *13*, 751 (1976)
102. Appelbaum, J. A., Baraff, G. A., Hamann, D. R.: Phys. Rev. *B14* (1976) 1623
103. Van Bommel, A. J., Crombeen, J. E.: Surf. Sci. *57*, 437 (1976)
104. Phillips, J. C.: Surf. Sci. *40*, 459 (1973)
105. Ludeke, R., Ley, L.: Proc. XIV. Int. Conf. Semicond. Phys., Edinburgh, 1978, p. 1069
106. Sakaki, H., Chang, L. L., Ludeke, R., Chang, C.-A., Sai-Halasz, G. A., Esaki, L.: Appl. Phys. Lett. *31*, 211 (1977)
107. Chang, C.-A., Ludeke, R., Chang, L. L., Esaki, L.: Appl. Phys. Lett. *31*, 759 (1977)
108. Benninghoven, A.: Surf. Sci. *53*, 569 (1975)
109. Dowsett, M. G., King, R. M., Parker, E. H. C.: Appl. Phys. Lett. *31*, 529 (1977)
110. Freeouf, J. L., Eastman, D. E.: CRC Crit. Rev. Solid State Sci. *5*, 245 (1975)
111. Larsen, P. K., Neave, J. H., Joyce, B. A.: Preprint (1979) submitted for publication to Phys. Rev. Lett.
112. Abstreiter, G., Pinczuk, A., Trommer, R., Tsu, R.: Proc. XIII Int. Conf. Semicond. Phys., Rome, 1976, p. 779
113. Abstreiter, G.: Verhandl. DPG VI, *12*, 81 (1977), and private communication (1978)
114. Barnes, P. A., Cho, A. Y.: Appl. Phys. Lett. *33*, 651 (1978)
115. ISA Riber technical brochure No. 265/Apr. 78; PHI data sheet 1057 9-78 15 M; V. G. data sheet 09-008-977
116. Günther, K. G.: In: Use of thin films in physical investigations, p. 213. J. C. Anderson ed. London: Academic 1966
117. Cho, A. Y.: J. Vac. Sci. Technol. *8*, S31 (1971)
118. Rode, D. L.: J. Crystal Growth *27*, 313 (1974)
119. Van Vechten, J. A.: J. Crystal Growth *38*, 139 (1977)
120. Joyce, B. A., Foxon, C. T.: Japan. J. Appl. Phys. *16*, Suppl. 16-1, 17 (1977)
121. Joyce, B. A., Foxon, C. T., Neave, J. H.: J. Japan. Assoc. Crystal Growth *5*, 185 (1978)

122. Tsang, W. T., Cho, A. Y.: Appl. Phys. Lett. *30,* 293 (1977)
123. Gowers, J. P., Joyce, B. A.: Private communication (1978)
124. Tsang, W. T.: Appl. Phys. Lett. *34,* 473 (1979)
125. Arthur, J. R.: Private communication reported in Ref.[55]
126. Kennedy, D. P., Murley, P. C., Kleinfelder, W.: IBM J. Res. Dev. *12,* 399 (1968)
127. Kennedy, D. P., O'Brien, R. R.: IBM J. Res. Dev. *13,* 212 (1969)
128. McLevige, M. V., Vaidyanathan, K. V., Streetman, B. G., Ilegems, M., Comas, J., Plew, L.: Appl. Phys. Lett. *33,* 127 (1978)
129. Ilegems, M., Dingle, R.: Proc. 5th Int. Symp. on GaAs and Related Compounds, Deauville, 1974, Inst. Phys. Conf. Ser. *24,* 1 (1975)
130. Arthur, J. R.: Surf. Sci. *38,* 394 (1973)
131. Laurence, G., Joyce, B. A., Foxon, C. T., Janssen, A. P., Samuel, G. S., Venables, J. A.: Surf. Sci. *68,* 190 (1977)
132. Naganuma, M., Takahashi, K.: Appl. Phys. Lett. *27,* 342 (1975)
133. Matsunaga, N., Suzuki, T., Takahashi, K.: J. Appl. Phys. *49,* 5710 (1978)
134. Bean, J. C.: J. Electrochem. Soc. *126,* 121C (1979), and private communication
135. Ploog, K., Fischer, A.: Unpublished
136. Wood, C. E. C.: Appl. Phys. Lett. *33,* 770 (1978)
137. Rode, D. L., Knight, S.: Phys. Rev. *B3,* 2534 (1971)
138. Rode, D. L.: In: Semiconductors and semimetals, Vol. 10, p. 1. R. K. Willardson and A. E. Beer, eds. New York: Academic 1975
139. Casey, H. C., Jr., Cho, A. Y., Barnes, P. A.: IEEE J. Quantum Electronics QE-*11,* 467 (1975)
140. Heckingbottom, R., Todd, C. J., Davies, G. J.: J. Electrochem. Soc. *126,* 121C (1979) and private communication
141. Cho, A. Y., Casey, H. C., Jr., Foy, P. W.: Appl. Phys. Lett. *30,* 397 (1977)
142. Fischer, A., Künzel, H., Ploog, K.: Verhandl. DPG (VI) *14,* 153 (1979)
143. Teramoto, I.: J. Phys. Chem. Solids *33,* 2089 (1972)
144. Van Vechten, J. A.: J. Electrochem. Soc. *122,* 423 (1975)
145. Lang, D. V., Logan, R. A., Kimerling, L. E.: Phys. Rev. *B15* (1977) 4874
146. Cho, A. Y., Hayashi, I.: Metall. Trans. *2,* 777 (1971)
147. Cho, A. Y.: Private communication (1978)
148. Van der Pauw, L. J.: Philips Res. Rep. *13,* 1 (1959)
149. Stillman, G. E., Wolfe, C. M.: Thin Solid Films *31,* 69 (1976)
150. Blood, P., Orton, J. W.: Rep. Prog. Phys. *41,* 157 (1978)
151. Wolfe, C. M., Stillman, G. E., Dimmock, J. O.: J. Appl. Phys. *41,* 504 (1970)
152. Brooks, H.: In: Advances in electronics and electron physics. Vol. 7, p. 85, New York: Academic 1955; and C. Herring, unpublished
153. Dingle, R., Störmer, H. L., Gossard, A. C., Wiegmann, W.: Appl. Phys. Lett. *33,* 667 (1978)
154. Abstreiter, G., Ploog, K.: Phys. Rev. Lett. *42,* 1308 (1979)
155. Störmer, H. L., Dingle, R., Gossard, A. C., Wiegmann, W., Sturge, M. D.: Solid State Commun. *29,* 705 (1979)
156. Tsui, D. C., Logan, R. A.: Appl. Phys. Lett., *35,* 99 (1979)
157. Wiley, J. D.: In: Semiconductors and semimetals, Vol. 10, p. 91. R. K. Willardson and A. C. Beer, eds. New York: Academic 1975
158. Fischer, A., Kamiya, T., Ploog, K.: unpublished
159. Calawa, A. R.: Appl. Phys. Lett. *33,* 1021 (1978)
160. Hilibrand, J., Gold, R. D.: RCA Rev. *21,* 245 (1960)
161. Copeland, J. A.: IEEE Trans. Electron Devices ED-*17,* 404 (1970)
162. Miller, G. L.: IEEE Trans. Electron Devices ED-*19,* 1103 (1972)
163. Künzel, H., Graf, K., Hafendörfer, M., Fischer, A., Ploog, K.: Technisches Messen, to be published
164. Binet, M.: Electron. Lett. *11,* 580 (1975)
165. Cho, A. Y., Dernier, P. D.: J. Appl. Phys. *49,* 3328 (1978)
166. Sze, S. M.: Physics of semiconductor devices, New York: Wiley 1969
167. Milness, A. G., Feucht, D. L.: Heterojunctions and metal semiconductor junctions, New York: Academic 1972

168. Massies, J., Devoldere, P., Etienne, P., Linh, N. T.: Proc. 7th Int. Vac. Congr. 3rd Int. Conf. Solid Surf. (Vienna, 1977) p. 639
169. Massies, J., Devoldere, P., Linh, N. T.: J. Vac. Sci. Technol. *15,* 1353 (1978)
170. Pinczuk, A., Abstreiter, G., Trommer, R., Cardona, M.: Solid State Commun. *21,* 959 (1977)
171. Tsu, R., Kawamura, H., Esaki, L.: Proc. XI. Int. Conf. Semicond. Phys., Warsaw, 1972, Vol. *2,* p. 1135
172. Abstreiter, G.: Private communication (1978)
173. Pinczuk, A., Worlock, J. M., Nahory, R. E., Pollack, M. A.: Appl. Phys. Lett. *33,* 461 (1978)
174. Bebb, H. B., Williams, E. W.: In: Semiconductors and semimetals, Vol. 8, p. 181 and 321. R. K. Willardson and A. C. Beer, eds. New York: Academic, 1972
175. Ashen, D. J., Dean, P. J., Hurle, D. T. J., Mullin, J. B., White, A. M., Greene, P. D.: J. Phys. Chem. Solids *36,* 1041 (1975)
176. Queisser, H. J.: Appl. Phys. *10,* 275 (1976)
177. Scott, C. B., Roberts, J. S.: Proc. 7th Int. Symp. on GaAs and Related Compounds, 1978, Inst. Phys. Conf. Ser. *45,* 181 (1979)
178. Ozeki, M., Nakai, K., Dazai, K., Ryuzan, O.: Japan. J. Appl. Phys. *13,* 1121 (1974)
179. This sample was kindly supplied by D. M. Collins, Corporate Solid State Laboratory, Varian Associates, Palo Alto, Ca. 94303, U.S.A.
180. Lang, D. V., Cho, A. Y., Gossard, A. C., Ilegems, M., Wiegmann, W.: J. Appl. Phys. *47,* 2558 (1976)
181. Koschel, W. H., Hiesinger, P.: Private communication (1979)
182. Morkoc, H., Cho, A. Y.: J. Electrochem. Soc. *126,* 121C (1979)
183. Meggitt, B. T., Parker, E. H. C., King, R. M.: Appl. Phys. Lett. *33,* 528 (1978)
184. Cho, A. Y., J. Vac. Sci. Technol. *16,* 275 (1979)
185. Cho, A. Y., Dunn, C. N., Kuvas, R. L., Schroeder, W. E.: Appl. Phys. Lett. *25,* 224 (1974)
186. Cho, A. Y., DiLorenzo, J. V., Mahoney, G. E.: IEEE Trans. Electron Devices ED-*24,* 1186 (1977)
187. Ballamy, W. C., Cho, A. Y.: IEEE Trans. Electron Devices ED-*23,* 481 (1976)
188. Schneider, M. V., Linke, R. A., Cho, A. Y.: Appl. Phys. Lett. *31,* 219 (1977)
189. Linke, R. A., Schneider, M. V., Cho, A. Y.: IEEE Trans. Microwave Theory Tech. MTT-*26,* 935 (1978)
190. Williams, R. E., Shaw, D. W.: Electronics Lett. *13,* 408 (1977)
191. Hewitt, B. S., Cox, H. M., Fukui, H., DiLorenzo, J. V., Schlosser, W. O., Iglesias, D. E.: Proc. 6th Int. Symp. on GaAs and Related Compounds, Edinburgh, 1976, Inst. Phys. Conf. Ser. *33a,* 246 (1977)
192. Cox, H. M., Hewitt, B. S., Wemple, S. H., DiLorenzo, J. V., Schlosser, W. O., Magalhaes, F. M.: Proc. 6th Int. Symp. on GaAs and Related Compounds, St. Louis, 1976, Inst. Phys. Conf. Ser. *33b,* 271 (1977)
193. Van der Ziel, J. P, Mikulyak, R. M., Cho, A. Y.: Appl. Phys. Lett. *27,* 71 (1975)
194. Merz, J. L., Cho, A. Y.: Appl. Phys. Lett. *28,* 456 (1976)
195. Merz, J. L., Gossard, A. C., Wiegmann, W.: Appl. Phys. Lett. *30,* 629 (1977)
196. Cho, A. Y., Yariv, A., Yeh, P.: Appl. Phys. Lett. *30,* 471 (1977)
197. Yeh, P., Yariv, A., Cho, A. Y.: Appl. Phys. Lett. *32,* 104 (1978)
198. Merz, J. L., Logan, R. A., Wiegmann, W., Gossard, A. C.: Appl. Phys. Lett. *26,* 337 (1975)
199. Reinhart, F. K., Cho, A. Y.: Appl. Phys. Lett. *31,* 457 (1977)
200. Reinhart, F. K., Logan, R. A.: Appl. Phys. Lett. *26,* 288 (1974)
201. Cho, A. Y., Casey, H. C., Jr.: Appl. Phys. Lett. *25,* 288 (1974)
202. Cho, A. Y., Dixon, R. W., Casey, H. C., Jr., Hartman, R. L.: Appl. Phys. Lett. *28,* 501 (1976)
203. Tsang, W. T.: Appl. Phys. Lett. *33,* 245 (1978)
204. Miller, B. I., McFee, J. H., Martin, R. J., Tien, P. K.: Appl. Phys. Lett. *33,* 44 (1978)
205. Lee, T. P., Cho, A. Y.: Appl. Phys. Lett. *29,* 164 (1976)
206. Casey, H. C., Jr., Somekh, S., Ilegems, M.: Appl. Phys. Lett. *27,* 142 (1975)
207. Lee, T. P., Holden, W. S., Cho, A. Y.: Appl. Phys. Lett. *32,* 415 (1978)
208. Ilegems, M., Schwartz, B., Koszi, L. A., Miller, R. C.: Appl. Phys. Lett. *33,* 629 (1978)
209. DiLorenzo, J. V., Niehaus, W. C., Cho, A. Y.: J. Appl. Phys. *50,* 951 (1979)

210. Devlin, J., Wood, C. E. C., Stall, R.: Paper presented on the 37th Ann. Device Res. Conf., Boulder, 1979
211. Tsang, W. T.: Appl. Phys. Lett. *33*, 1022 (1978)
212. Casey, H. C., Jr., Cho, A. Y., Nicollain, E. H.: Appl. Phys. Lett. *32*, 678 (1978)
213. Casey, H. C., Jr., Cho, A. Y., Lang, D. V., Nicollain, E. H.: J. Vac. Sci. Technol. *15*, 1408 (1978)
214. Casey, H. C., Jr., Cho, A. Y., Lang, D. V., Nicollain, E. H., Foy, P. W.: J. Appl. Phys. *50*, 3484 (1979)
215. Esaki, L., Tsu, R.: IBM J. Res. Dev. *14*, 61 (1970)
216. Cho, A. Y.: Appl. Phys. Lett. *19*, 467 (1971)
217. Chang, L. L., Esaki, L., Tsu, R.: Appl. Phys. Lett. *24*, 593 (1974)
218. Esaki, L., Chang, L. L.: Phys. Rev. Lett. *33*, 495 (1974)
219. Dingle, R., Wiegmann, W., Henry, C. H.: Phys. Rev. Lett. *33*, 827 (1974)
220. Dingle, R., Gossard, A. C., Wiegmann, W.: Phys. Rev. Lett. *33*, 1327 (1974)
221. Dingle, R.: In: "Festkörperprobleme" (Advances in Solid State Physics), Vol. XV, p. 21. H. J. Queisser, ed., Braunschweig: Pergamon/Vieweg 1975
222. Tsu, R., Chang, L. L., Sai-Halasz, G. A., Esaki, L.: Phys. Rev. Lett. *34*, 1509 (1975)
223. Manuel, P., Sai-Halasz, G. A., Chang, L. L., Chang, C.-A., Esaki, L.: Phys. Rev. Lett. *37*, 1701 (1976)
224. Sai-Halasz, G. A., Pinczuk, A., Yu, P. Y., Esaki, L.: Surf. Sci. *73*, 232 (1978)
225. Chang, L. L., Sakaki, H., Chang, C.-A., Esaki, L.: Phys. Rev. Lett. *38*, 1489 (1977)
226. Chang, L. L.: Surf. Sci. *73*, 226 (1978)
227. Dingle, R.: Surf. Sci. *73*, 229 (1978)
228. Sai-Halasz, G. A., Tsu, R., Esaki, L.: Appl. Phys. Lett. *30*, 651 (1977)
229. Sai-Halasz, G. A., Chang, L. L., Wolter, J. M., Chang, C.-A., Esaki, L.: Solid State Commun. *26*, 935 (1978)
230. Sakaki, H., Chang, L. L., Sai-Halasz, G. A., Chang, C.-A., Esaki, L.: Solid State Commun. *26*, 589 (1978)
231. Sai-Halasz, G. A., Esaki, L., Harrison, W. A.: Phys. Rev. *18*, 2812 (1978)
232. Chang, L. L., Sai-Halasz, G. A., Kawai, N. J., Esaki, L.: J. Vac. Sci. Technol. *16* (1979)
233. Chang, L. L., Segmüller, A., Esaki, L.: Appl. Phys. Lett. *28*, 39 (1976)
234. Dingle, R., Gossard, A. C., Wiegmann, W.: Bull. Am. Phys. Soc. *21*, 367 (1976)
235. Fleming, R. M., McWhen, D. B., Gossard, A. C., Wiegmann, W.: Bull. Am. Phys. Soc. *24*, 249 (1979)
236. Gossard, A. C.: Proc. 4th Int. Thin Films Congr. (1978), Thin Solid Films *57*, 3 (1979)
237. Petroff, P. M.: J. Vac. Sci. Technol. *14*, 973 (1977)
238. Schulman, J. N., McGill, T. C.: Phys. Rev. Lett. *39*, 1680 (1977)
239. Caruthers, E., Lin-Chung, P. J.: Phys. Rev. B*17*, 2705 (1978)
240. Andreoni, W., Baldereschi, A., Car, R.: Bull. Am. Phys. Soc. *23*, 292 (1978)
241. Van der Ziel, J. P., Gossard, A. C.: J. Appl. Phys. *48*, 3018 (1977)
242. Van der Ziel, J. P., Gossard, A. C.: Phys. Rev. B*17*, 765 (1977)
243. Merz, J. L., Barker, A. S., Jr., Gossard, A. C.: Appl. Phys. Lett. *31*, 117 (1977)
244. Barker, A. S., Jr., Merz, J. L., Gossard, A. C.: Phys. Rev. B*17*, 3181 (1978)
245. Petroff, P. M., Gossard, A. C., Savage, A., Wiegmann, W.: J. Crystal Growth, *46*, 172 (1979)
246. Petroff, P. M., Gossard, A. C., Savage, A., Wiegmann, W.: to be published (1979)
247. Gossard, A. C.: To be published
248. Abstreiter, G., Ploog, K.: To be published
249. Abstreiter, G., Ploog, K., Fischer, A.: Germ. Pat. Appl. (1979) P 2913068.5
250. Wataze, M., Mitsui, Y., Shimanoe, T., Nakatani, M., Mitsui, S.: Electron. Lett. *14*, 759 (1978)
251. Dupuis, R. D., Dapkus, P. D., Chin, R., Holonyak, N., Jr., Kirchoefer, S. W.: Appl. Phys. Lett. *34*, 265 (1978)
252. Dupuis, R. D., Dapkus, P. D., Garner, C. M., Su, C. Y., Spicer, W. E.: Appl. Phys. Lett. *34*, 335 (1979)

Received June 11, 1979

Author Index Volumes 1–3

The volume number are printed in italics

A. F. Williams

A Theoretical Approach to Inorganic Chemistry

1979. 144 figures, 17 tables. XII, 316 pages
ISBN 3-540-09073-8

This book outlines the application of simple quantum mechanics to the study of inorganic chemistry, and shows its potential for systematizing and understanding the structure, physical properties, and reactivities of inorganic compounds. The considerable strides made in inorganic chemistry in recent years necessitate the establishment of a theoretical framework if the student is to acquire a sound knowledge of the subject. A wide range of topics is covered, and the reader is encouraged to look for further extensions of the theories discussed. The book emphasizes the importance of the critical application of theory and, although it is chiefly concerned with molecular orbital theory, other approaches are discussed. This text is intended for students in the latter half of their undergraduate studies. (235 references)

From the Foreword by Prof. C. K. Jørgensen

"... Dr. Alan Williams has acquired a considerable experience in work with transition metal complexes at the Universities of Cambridge and Geneva. In this book he has tried to avoid the variety of ephemeral and often contradictory rationalisations encountered in this field, and has made a careful comparison of modern opinions about chemical bonding. In my opinion this effort is fruitful for all students and active scientists in the field of inorganic chemistry. The distant relations to group theory, atomic spectroscopy and epistemology are brought into daylight.
... The interdisciplinary approach of the book shows up in the careful consideration given to many experimental techniques such as vibrational (infra-red and Raman), electronic (visible and ultraviolet), Mössbauer, magnetic resonance, and photoelectron spectra, with data for gaseous and solid samples as well as selected facts about solution chemistry. The book could not have been written a few years ago, and is likely to remain a highly informative survey of modern inorganic chemistry and chemical physics."

G. Meier, E. Sackmann, J. G. Grabmaier

Applications of Liquid Crystals

1975. 87 figures, 11 tables. V, 164 pages
ISBN 3-540-07302-7

"As interest in liquid crystals has increased over the past ten years, considerable progress has been made in this field. Physicists, chemists and those in manufacturing an processing industries will find this concise presentation of the theory and possible applications of liquid crystals most useful.
The book is divided into three parts and the first, by Meier, provides an introduction to the physics of liquid crystals. He discusses the structures and properties of the three basic types of liquid crystals – smectic, nematic and cholesteric, and also considers electric field effects. Liquid crystals are excellent solvents for organic molecules and in the second chapter Sackmann deals with the potential applications of liquid crystals in fundamental research. He discusses their use in magnetic resonance spectroscopy, optical spectroscopy, gas – liquid chromatography, and as anisotropic solvents for the determination of molecular properties. The final part of the book, by Grabmaier, is devoted to the medical and technical applications..."

J. Mol. Structure

"This is a timely book ...
This volume is well illustrated with line drawings and black and white photographs...
In the preface the authors of this book make the claim that it will appeal not only to physicists and chemists (especially spectroscopists) but equally to those in the manufacturing and processing industries (including electrical engineers). And they are right."

J. Applied Crystallography

Springer-Verlag
Berlin
Heidelberg
New York

MIX
Papier aus verantwortungsvollen Quellen
Paper from responsible sources
FSC® C105338

If you have any concerns about our products,
you can contact us on
ProductSafety@springernature.com

In case Publisher is established outside the EU,
the EU authorized representative is:
Springer Nature Customer Service Center GmbH
Europaplatz 3, 69115 Heidelberg, Germany

Printed by Libri Plureos GmbH
in Hamburg, Germany